MONOGRAPHS ON
APPLIED PROBABILITY AND STATISTICS

General Editor

D. R. COX, F. R. S.

IDENTIFICATION OF OUTLIERS

Identification of Outliers

D. M. HAWKINS
Senior Consultant in Operations Research and Statistics,
Council for Scientific and Industrial Research, South Africa

1980

SPRINGER-SCIENCE+BUSINESS MEDIA, B.V.

150TH ANNIVERSARY

Originally published by Chapman & Hall in 1980

British Library Cataloguing in Publication Data

Hawkins, D M
Identification of outliers. – (Monographs on applied probability and statistics).
1. Outliers (Statistics)
I. Title II. Series
519.5 QA276

ISBN 978-94-015-3996-8 ISBN 978-94-015-3994-4 (eBook)
DOI 10.1007/978-94-015-3994-4

Contents

Preface

The problem of outliers is one of the oldest in statistics, and during the last century and a half interest in it has waxed and waned several times. Currently it is once again an active research area after some years of relative neglect, and recent work has solved a number of old problems in outlier theory, and identified new ones. The major results are, however, scattered amongst many journal articles, and for some time there has been a clear need to bring them together in one place. That was the original intention of this monograph: but during execution it became clear that the existing theory of outliers was deficient in several areas, and so the monograph also contains a number of new results and conjectures.

In view of the enormous volume of literature on the outlier problem and its cousins, no attempt has been made to make the coverage exhaustive. The material is concerned almost entirely with the use of outlier tests that are known (or may reasonably be expected) to be optimal in some way. Such topics as robust estimation are largely ignored, being covered more adequately in other sources. The numerous *ad hoc* statistics proposed in the early work on the grounds of intuitive appeal or computational simplicity also are not discussed in any detail.

The main emphasis of the monograph is on the salient theoretical aspects of the problem, but the appendices of fractiles of many of the tests discussed will also be of use to practitioners interested in applying outlier tests to data.

I wish to thank the Universities of the Witwatersrand and North Carolina – the former for granting a year's sabbatical leave in which to write this monograph, and the latter for providing facilities for preparing it and carrying out some of the original research it embodies. The final preparation was carried out under the aegis of the Council for Scientific and Industrial Research, and all three bodies have provided me with invaluable facilities and considerable amounts of computer time for the computation of tables.

I am grateful to the publishers of the following journals for permission to use copyright material from their pages:

Utilitas Mathematica for Appendix 1;

Journal of Statistical Computation and Simulation for Appendix 3;

Statistica Neerlandica for Appendix 5;

Annals of Mathematical Statistics for Appendix 6.

I also thank those colleagues who helped with the preparation of the manuscript or contributed ideas, in particular Drs R. Carroll and D. Bradu for helpful discussions of early drafts, Mrs J. Galpin for the computation of the tables in Appendices 8 and 10, and Miss S. van den Berg for typing special symbols into the manuscript.

Finally, I acknowledge gratefully the help of my wife, Greer, who prepared most of the manuscript for computer formatting, helped with the running of programs and cheerfully tolerated the disruption of her life caused by the monograph.

Pretoria, March 1979. D. M. Hawkins

CHAPTER 1

Introduction

1.1 What is an outlier?

Any applied statistician who has analysed a number of sets of real data is likely to have come across 'outliers'. The intuitive definition of an outlier would be 'an observation which deviates so much from other observations as to arouse suspicions that it was generated by a different mechanism'. An inspection of a sample containing outliers would show up such characteristics as large gaps between 'outlying' and 'inlying' observations and the deviation between the outliers and the group of inliers, as measured on some suitably standardized scale.

1.2 Genesis of outliers

There are two basic mechanisms which give rise to samples which appear to have outliers. It is a matter of some importance which of the mechanisms generated any particular set of observations since this consideration certainly affects, or should affect, one's subsequent analysis of the data.

Mechanism (*i*): The data come from some heavy tailed distribution such as Student's t. There is no question that any observation is in any way erroneous.

Formalizing this model, Green (1976) has introduced a classification of families of statistical distributions into those that are 'outlier-prone' and those that are 'outlier-resistant'. The 'outlier-prone' families have tails which go to zero slowly: a distribution is said to be absolutely outlier-prone if (letting $X_{n,i}$ be the ith order statistic based on a sample of size n) there exists ε, $\delta > 0$ and an integer n_0 such that

$$\Pr[X_{n,n} - X_{n,n-1} > \varepsilon] \geqslant \delta \quad \text{for all} \quad n > n_0 \qquad (1.1)$$

and is relatively outlier-prone if there exist $c > 1$, $\delta > 0$ and n_0 such that

$$\Pr[X_{n,n}/X_{n,n-1} > c] \geqslant \delta \quad \text{for all} \quad n > n_0 \tag{1.2}$$

Clearly if either of these situations holds, then there will be a tendency for the largest order statistic to be suspiciously large relative to its predecessor, and so samples generated by outlier-prone distributions will tend to contain visual outliers.

Absolutely and relatively outlier-resistant distributions are those which are not absolutely and relatively outlier-prone respectively. The former class includes the normal family of distributions and the latter the gamma family.

Mechanism (ii): The data arise from two distributions. One of these, the 'basic distribution', generates 'good' observations, while another, the 'contaminating distribution', generates 'contaminants'. If the contaminating distribution has tails which are heavier than those of the basic distribution, then there will be a tendency for the contaminants to be outliers – that is, to separate visibly from the good observations, which will then constitute the inliers.

When this mechanism is appropriate, it may often be invoked in either of two ways. The first, *mechanism (iia)*, specifies that in the sample of size n, exactly $n-k$ observations come from the basic distribution, and k from the contaminating distribution. For this model to be of use presupposes a knowledge of k – the number of contaminants in the sample.

More commonly, k is not known. A suitable model here is the Dixon (1950) *mechanism (iib)*: with probability p, any given observation comes from the contaminating distribution, while with probability $1-p$ it comes from the basic distribution. The value of k is then a random variable following a binomial distribution. In most cases of interest, p is assumed to be near zero, even if its actual value is not known.

We may note that if the basic distribution is $f_0(.)$ and the contaminating distribution is $f_1(.)$ then the distribution of an arbitrary observation is

$$f(.) = (1-p)f_0(.) + pf_1(.) \tag{1.3}$$

Viewed in this way, the mechanism iib thus reduces to a particular case of mechanism (i), the heavy-tailed distribution model. It is, however, a case of especial interest, since problems of estimation will generally involve the individual parameters of the distributions $f_0(.)$

and $f_1(.)$, rather than those of $f(.)$. Thus it is well to maintain the distinction between the two models.

1.3 Treatment of outliers

1.3.1 *Estimation*

The effect of outliers on the analysis of a set of data depends strongly on the mechanism by which the outliers are believed to be generated. If mechanism (i) is assumed, then the outliers, despite appearances, are valid observations from the distribution under study. Usually, the major objective of the analysis will be to estimate a parameter – for example location – of this distribution. If the form of the distribution is known (for example, if it is a Cauchy distribution), then this estimation poses no especial difficulties – one uses whatever estimator is best for that particular distributional form.

If the distributional form is not known, then one would be tempted to use a robust estimator. To estimate location, for example, one might use a trimmed mean – the mean of the central order statistics of the sample. While the trimmed mean is an optimal estimator of the location of a certain, highly specific distributional family, it is an estimator that is obviously good under very broad conditions – those in which a heavy tail gives rise to some outliers, while the central order statistics behave like a sample from the normal distribution. This is likely to be at least approximately true for most symmetric unimodal distributions with heavy tails.

We may note that the use of a trimmed mean using, say, the central $n - k$ values consists of discarding the $k/2$ most extreme observations on each side, regardless of their magnitude, and finding the mean of those left. Contrast this procedure with the following 'rejection of outliers' procedure – 'discard all those extreme order statistics whose deviation from the remainder of the sample is sufficiently large. Use as location estimator the arithmetic mean of the data retained.'

This comparison shows that the 'rejection of outliers' approach may be regarded as a method of obtaining a robust estimator. It has the advantage over the trimmed mean that if the sample is truly normal, then the fully efficient arithmetic mean of all observations will often be used as the estimator. Thus it is very plausible that, with suitable selection of a procedure for deciding whether the deviations of the extreme values are sufficiently large, the 'rejection of outliers' procedure can be made to give estimators which are both robust for distributions with heavier tails than was supposed; and efficient, in that, if the distribution assumed is correct, use is frequently made of the fully efficient estimator for that distribution.

The idea of using rejection as a method of obtaining robust estimators was set out formally by Anscombe (1960). Using a mixture model, mechanism (iib), he set up a rejection rule in terms of an 'insurance policy': As a result of being willing to occasionally reject a good observation, one incurs a cost in that the mean squared error of the resulting estimate applied to uncontaminated samples will be larger than would otherwise be the case. If we regard this increase as being like an insurance premium, then the protection that it buys is a decrease in mean squared error when the sample is actually contaminated. Some simulation studies have shown that a large amount of protection may be bought for a small premium.

While these comments apply in quite general terms, there is an especially compelling interest in the rejection approach if mechanism (ii) (in either of its variants) is believed to hold, for then, one can regard rejection as a method of partitioning the data from $f_0(.)$ and those from $f_1(.)$ prior to using the partitioned data to estimate the parameters of $f_0(.)$ and $f_1(.)$.

In the multivariate context, of course, mechanism (iib), mixtures of distributions, is familiar as the mathematical model underlying many problems in cluster analysis (see for example Hartigan, 1975). The standard methods of cluster analysis include two broad classes: one seeks to estimate p and the parameters of $f_0(.)$ and $f_1(.)$ from the entire sample, after which a discriminant analysis may be used to partition the data; the other starts by finding an optimal partitioning of the data – from this the parameters may be estimated. The similarity between the latter approach and rejection of outliers is unmistakable.

There is another area in which the use of rejection of outliers represents a more viable approach to robust estimation, and that is the situation in which the data are not identically distributed. A good example of this is multiple regression. While robust multiple regression procedures exist, they are still rather awkward computationally. The approach of eliminating several outliers and estimating on the basis of the remainder of the sample involves quite simple matrix operations, and can lead to considerable insight into the structure of the underlying process (see Chapter 7).

1.3.2 *Smoothly varying weights*

A valid objection to the outlier rejection approach is the influence on the final estimator of an individual observation. Take, for example, the location estimator consisting of the arithmetic mean of the observations retained. The derivative of this quantity with respect to a particular observation is a constant until that observation reaches the cutoff point at which it is rejected; the derivative then drops at once to zero. The undesirability of this is clear, and more than a century ago, Glaisher (1872) proposed an alternative. His suggestion is that the location of the underlying distribution be estimated by a weighted mean of the data. The weights are computed iteratively, and follow from a Bayesian model in which the observations have different standard deviations, but are normally distributed with common mean.

This proposal seems to be the ancestor of Bayesian approaches suggested more recently (see Chapter 9) for estimating the parameters of the basic distribution in the face of outlier contamination.

Another approach to estimation is the use of Winsorization. The idea underlying this is that all observations which are regarded as excessively outlying are moved to a prefixed position near the central observations. In this way, the outlying observations are not discarded entirely in estimating the parameters of interest.

When deciding whether to use Winsorization or outright rejection of outlying observations, one should be guided by the underlying model of the data. If the observations are generated by mechanism (i), the heavy-tailed distribution, and one wishes to estimate the parameters of this distribution, then the outliers represent valid observations. Thus one should be reluctant to discard them entirely, and hence prefer to use Winsorization, which is robust, but does make partial use of the outliers.

If, on the other hand, one believes that mechanism (ii) is operative, and one is interested in estimating the parameters of the basic distribution, then Winsorization should not be used. In this case, the outliers may be presumed to come from the contaminating distribution, and hence to contain no information about the basic distribution. Thus to the extent that one is sure they are not from the basic distribution, one should ignore them. This may be done in a classical way by deleting them as soon as one rejects the null hypothesis that they come from the basic distribution, or in a Bayesian way by assigning them smaller weights as they deviate more from the basic distribution.

Of course this distinction, quite clear in theory, may become somewhat blurred in practice – for example, when the basic and the contaminating distributions are believed to share a common parameter. If this is the case, then it is no longer optimal to discard contaminants, but the manner in which they are incorporated into the estimation will depend on the specific distributions in the model. In the light of this comment it is of interest to note that many studies of the performance of estimators for contaminated samples have assumed a parameter in common between the two distributions.

1.3.3 *'Sounding a warning'*

Another possible use of a procedure for identifying outliers is to provide a check on model assumptions. For example, if one wishes to carry out some parametric analysis assuming normality of data, then various checks will be applied to the data to ensure that the model fits well enough. One such test could well be an outlier test, and such a test will be sensitive to the tail behaviour of the true distribution. This may be an advantage if the parametric analysis is sensitive to departures from model is the tails of the distribution, but not elsewhere. This is the case for many techniques (for example analysis of variance) where skewness and high kurtosis are the most damaging departures from model.

As the phrase 'sounding a warning' implies, this use of an outlier identification procedure is made in much the same way as a preliminary test for homogeneity of variance in an analysis of variance. The test is regarded as a screen – if the data pass, then the standard parametric procedure will be applied; otherwise some other, unspecified, action will be taken. The range of actions includes, but is by no means restricted to, removing any outliers from the sample and carrying out the analysis proposed originally on the remaining 'clean' observations.

The fact that any outlier test may be used to test for departures from model has a corollary: since the presence of outlier is indicative of some departure from the model assumed for the data, a goodness-of-fit test statistic may, in principle, be used to test for the presence of outliers. Two goodness-of-fit tests which are very effective for detecting the presence of outliers in a normal sample are the moment coefficients of skewness and kurtosis, and both in fact are optimal in a particular sense.

A more general class of goodness-of-fit tests is that based on the sample distribution function. Provided all parameters are known (or can be estimated from a sufficiently large sample) the probability integral transformation may be used to bring this to a standard form.

The Anderson–Darling test, which is especially sensitive to departures from model in the tails of the distribution, has great intuitive appeal as an outlier test statistic. Its use has not, apparently, been proposed in the literature, but it is to be hoped that future work will cover its performance thoroughly.

1.3.4 *Classification of observations*

Finally, as noted earlier, when the data are generated by mechanism (ii), one may be interested in first testing whether any observations are contaminants, and if so, how many and which observations they are. This basic framework supports much of the theory of optimal outlier tests.

The supposition underlying the desire to classify the observations is that the contaminants are not an unmitigated nuisance whose ill effects on the sample are to be eliminated somehow; rather it is of interest to identify the observations from the basic and the contaminating distributions, and hence make deductions about the generating mechanism.

It is common knowledge that a number of important discoveries in astronomy have been made as a result of a careful study of sets of residuals from fitted orbits. While the author is not aware of any important effect that has shown up in the form of a few outliers (as opposed to a systematic lack of fit) it is nevertheless quite conceivable that this could occur. In such a case, the effect would be detected by: (a) testing for the presence of outliers in the sample, and then (b) locating and identifying the outliers, and searching for a possible explanation for them. This procedure is precisely the classification procedure set out above.

1.4 Slippage tests

A topic related quite closely to outlier theory is slippage testing. The usual model for this is that one is given observations X_{ij}, $j = 1$ to m_i, $i = 1$ to n. All are independently distributed, X_{ij} coming from a distribution $f_i(.)$. The slippage hypothesis is that while the majority of the $f_i(.)$ are identical to some common distribution $f(.)$, some small number of them have large probabilities in the tail regions of $f(.)$, and are said to have 'slipped'.

When one inspects a sample generated by a slippage hypothesis, one tends to find clusters of outliers, these outliers being the samples from the slipped population or populations. Of course if all m_i equal 1, then the slippage problem becomes formally identical to an outlier problem. Thus the slippage problem may be regarded quite usefully as a generalization of the outlier problem, and it is no surprise to learn that the major portion of the theory is common to both.

A slippage model is often a good one to use in an analysis of variance in which a large number of possible treatments are being screened to see whether some small number are markedly more effective than the rest. In such a case, the use of a slippage test will give a more powerful analysis than the usual F-test of analysis of variance. This extra power is a consequence of the use of a relatively restricted alternative hypothesis, and may be quite substantial (see Chapter 5). We may note a similarity between the use of a slippage test here and the 'ranking and selection' problem. The latter problem, starting from a statement that not all the $f_i(.)$ are identical, seeks to isolate the 'best treatment' – typically the stochastically largest $f_i(.)$.

1.5 The meaning of significance levels

Before the computer era, data analysts undoubtedly had much more contact with the observations they were processing and probably made more use of simple displays of data and grouping of data for the computation of sample statistics. In the process of doing this, they would become aware of whether the sample seemed to contain outliers, and if so, how many. It is also probably true that no sample, all of whose members appeared to be good, would ever be subjected to a formal outlier test.

Collett and Lewis (1976) point this fact out, and suggest as a consequence that the notion of the significance level of a formal outlier test is meaningless, since the test is unlikely ever to be applied unless outliers are quite obviously present.

In support of their contention, they present the results of a study in which analysts were presented with samples containing outliers of varying degrees of discordancy, and were asked to decide whether or not the samples contained outliers. The study showed clearly that this perception was affected by such irrelevant factors as the scale of the observations, the method of presentation and the value of ancillary statistics.

An objection may be made to this study that the subjects were apparently not skilled in statistical analysis. It is this author's experience that statisticians tend to detect outliers that are not present, and to regard the non-significance of outlier test statistics as a reflection on the poor power of the test rather than an indication that the suspicious-looking observation is statistically quite plausible.

A degree of support for the view that statisticians are quite good at locating outliers visually is provided by Relles and Rogers (1977). They provided experienced statisticians with random samples from several heavy-tailed t-distributions and asked them to provide estimates of location. The results showed that rejecting the observations considered by the statisticians to be outliers, and finding the mean of the remainder, produced quite good estimators of location. Somewhat counter to this is the observation that rejecting a further observation on each side yields a slightly better estimator.

A more pressing problem with regard to outliers, however, is that now being felt with computerized data bases. While the statistician plotting data values on graph paper may be quite good at finding outliers, the numbers in large data bases are seldom displayed, and so outliers run a real risk of remaining undetected because unsuspected. For this reason, fully automated outlier detection procedures should be an integral part of any large data base maintenance system, and all computer-based files of data should be checked for outliers before use. In this way, one can hope to trap anomalous cases, as well as recording blunders, and so ensure the integrity of the data base.

If this procedure is followed then it is meaningful to discuss the significance levels of an outlier test, since all samples are tested for outliers, and not just those which almost certainly contain outliers.

Further, if we believe that the average data analyst who would use an outlier test at all, will apply one whenever a sample appears at all suspicious, then once again, the concept of the significance level is meaningful as a statement applying to all sets of data seen by that analyst. This is because the samples which are not screened at all, would not yield a significance if they were tested for outliers.

1.6 Brief early history of outlier rejection

Surveys of the 18th- and 19th-century literature of statistics have shown that awareness of the outlier problem arose very early. The genesis of research into outlier rejection is bound up with that of the method of least squares, and the very extensive history of least squares given by Harter (1974, 1975, 1976) contains a full discussion of early writings on outlier rejection. Another account of the literature up to 1931 is to be found in Rider (1933), and so it is appropriate to outline only a few major discoveries and concepts, up to the major paper by Pearson and Chandra Sekar (1936).

No doubt the practice of rejecting observations which appeared to be discordant with other data has so much intuitive appeal that it was never 'discovered' formally, but merely came into use, possibly spreading by a process of osmosis. In 1778, we find Daniel Bernoulli commenting unfavourably on the practice (already widespread amongst astronomers of his time) of discarding any values that seemed discordant, and behaving as if the remaining observations constituted the entire sample.

In these early uses of outlier rejection, no formal statistical criterion was used – the rejection of outliers was made on the basis of the analyst's opinion of whether the sample values were mutually consonant. (The recent results of Relles and Rogers suggest that they may have done a better job in this than they are generally credited with.)

The controversy about whether or not one should reject outliers was not solved at once, and in fact remains an issue two centuries later. However, the subjective nature of outlier rejection was ameliorated in 1852 by Peirce's proposal of an objective criterion for the rejection of outliers. His criterion is based on the model that a mixture of distributions with unknown p generated the data. By making a worst-case assumption on p and supposing that the underlying basic distribution is $N(0, \sigma^2)$ with σ known, he sets up what is essentially a likelihood ratio test. Any residuals exceeding $c\sigma$ in absolute value are rejected where c is found as follows:

Let $\Phi(.)$ denote the cumulative distribution function of the standard normal distribution. Let s and s' denote the standard deviations of the full sample, and the sample stripped of N outliers, leaving $n' = n - N$ observations retained. Then c solves the equation

$$(s'/s)^{n'}[\{\exp\tfrac{1}{2}(c^2 - 1)\}2\Phi(-c)]^n = [N^N n'^{n'}/n^n]^{n/N} \qquad (1.4)$$

A number of writers raised theoretical and practical objections to this criterion, and Chauvenet (1866) proposed a different one. Noting that the expected number of observations in a sample of size n from $N(0, \sigma^2)$ exceeding $c\sigma$ in absolute value is $n\Phi(-c)$, he proposed that any residuals exceeding $c\sigma$ in absolute value be rejected, where c satisfies

$$n\Phi(-c) = 0.5 \qquad (1.5)$$

Thus Chavenet's criterion is so set as to reject, on average, half an observation of good data per sample, regardless of n.

This aspect of Chauvenet's test is especially significant, since it focuses on the experimentwise rejection rate, and not on the proportion of data values rejected. As n increases, so does c, and so the proportion of observations rejected decreases. It is easy by hindsight to see in this idea the kernel of the concept of simultaneous statistical inference, and indeed the test is not far from one with a preset experimentwise significance level.

By contrast, Stone (1868) proposed that any observer has a probability of $1/m$ of making a blunder or any particular observation, and proposed that c be chosen by

$$m\Phi(-c) = 0.5 \tag{1.6}$$

The apparent similarity between this and Chauvenet's rule is deceptive. Stone's rule rejects a fixed proportion of good observations, and so the number of outliers increases in proportion to the sample size.

A variant approach due to Irwin (1925) provided the basis for a single test for several outliers in the sample. Considering again a set of data distributed as $N(\xi^2, \sigma^2)$, σ known, Irwin considered the order statistics $X_{(1)} < X_{(2)} < \ldots < X_{(n)}$, and derived the distribution of the kth gap $(X_{(n-k+1)} - X_{(n-k)})/\sigma$ as a rather complicated infinite series. He also derived computationally tractable expressions for the moments of this kth gap, and showed that for $k = 1$ and 2, the distribution of the gap is well approximated by normal distributions whose means and variances he tabulated for several values of n. From this basic idea has sprung the modern work on gap tests.

All these criteria referred to the situation of observations distributed as $N(\xi, \sigma^2)$ with both ξ and σ known. In practice, of course, $\bar{X}$ was substituted for ξ, and s for σ, but the effect of this studentization depended on theory not yet extant.

Goodwin (1913) proposed that instead of the criterion $(X_i - \bar{X})/s$ used by earlier authors, one should substitute for $\bar{X}$ and s, the mean and standard deviation obtained omitting the outlier from the sample. Some years were to pass before it was shown that this test statistic is a monotonic function of $(X_i - \bar{X})/s$. The latter discovery by Thompson (1935) led to his derivation of the null distribution of an arbitrary studentized residual $(X_i - \bar{X})/s$. From this, he was able to deduce an outlier screening procedure that would reject a fixed proportion of all good data values and to produce a table of suitable critical values.

The final paper discussed in this short history is the 1936 paper of Pearson and Chandra Sekar. Using Thompson's distribution and some mathematical lemmas on extrema of the set of studentized residuals $(X_i - \bar{X})/s$, they showed that for sufficiently large c, the event $|X_i - \bar{X}|/s > c$ implies that for all $j \neq i$, $|X_j - \bar{X}|/s < c$. Thus if c is chosen to be the $\alpha/2n$ fractile of the distribution of $(X_i - \bar{X})/s$, then an outlier test rejecting any X_i satisfying $|X_i - \bar{X}|/s > c$ has an experimentwise significance level of α. Thus, in contrast to Thompson's work, which was in the spirit of Stone, theirs was in the spirit of Chauvenet.

This extremely significant paper also pointed out the existence of what has come to be termed the 'masking effect'. This effect arises when a sample contains more than one outlier. These outliers so increase the spread of the sample that the removal of a single outlier makes little improvement in the appearance of the sample, and in particular, all values of $(X_i - X)/s$ are near zero because of the very large value of s. The practical consequences of the masking effect is that any attempt to remove these outliers one at a time will prove fruitless, and so there is a need for more sophisticated methods that will detect all outliers.

Despite the shortness of this list of papers, they provide the basic ideas from which the subsequent work is derived. The details of this subsequent development will be sketched further in the chapters on various specific aspects of, and problems in outlier theory.

CHAPTER 2

General theoretical principles

2.1 Measures of performance

In Chapter 1, a distinction was made between situations in which our primary concern is to estimate one or more parameters of the underlying distribution, and that in which we wish to classify the data into 'good' data and outliers (if any).

Concentrating on the second of these situations then raises the question of how effective an outlier test is in doing its job. David and Paulson (1965) list five measures of performance of a test for a single outlier. Two of these are not applicable in complete generality and so will not be considered here. The remainder are:

(i) β_1, the probability of a correct decision, that is, the probability that the test concludes correctly that there is an outlier, and identifies it correctly.

(ii) β_2, the power function. This is the probability that the test concludes that there is an outlier in the sample, regardless of whether or not it identifies the right observation.

The difference between β_1 and β_2 is that the actual contaminant may be one of the central values in the sample, and the most extreme observation may be good. In this case, a significant value of the outlier statistic would warn us correctly that an outlier was present. If, however, we acted on it and rejected the most extreme observation then we would make an error (a type III error, in standard terminology).

These two measures are the prime contenders for our attention. The third measure is:

(iii) β_3, the conditional power – that is, the probability that the test is significant given that the contaminant is the extreme value tested. Clearly,

$$\beta_3 = \beta_1/\Pr(\text{contaminant is extreme})$$

The appeal of this measure lies in reasoning that if the contaminant is actually not the most extreme value of the sample then it is immaterial whether or not the test is significant. This measure of performance (and presumably the reasoning justifying it) has not been used to any great extent.

Not all measures exist for all tests. For example, the sample skewness may tell one that the sample contains one or more outliers, but it does not identify any specific observation as an outlier. Thus the measure β_1 and β_3 are meaningless for this test, but β_2 is perfectly well defined.

Clearly, when all measures exist, they can be expected to have very similar values when the distribution of contaminants differs markedly from that of good data, for then the contaminant is increasingly sure to be the most extreme value in the sample.

As the contaminating distribution approaches the true distribution, the three measures have the following limits for most tests symmetric in the observations

$$\beta_1 \to \alpha/n$$
$$\beta_2, \beta_3 \to \alpha$$

where α is the size of the test.

A number of theoretical results about the optimality of various tests have been deduced from these measures and the major ones will be sketched below.

2.2 Specific optimal tests

The measure of performance β_1 implies a statistic which may be used as the basis for a decision about whether or not a particular observation is an outlier. Thus a statistic has optimal β_1 if it is an optimal solution to the multiple decision problem of deciding between $D_0, D_1, \ldots, D_n$, where D_0 is the decision that no outliers are present, and D_i the decision that X_i is an outlier. Thus the work on β_1 optimal tests has generally approached the problem via decision theory.

In some theory of very broad applicability, Karlin and Truax (1960) consider the following problem. Suppose that for $i = 1$ to n, X_i has density $f(.|\theta_i)$ where the parameter θ_i is an unknown scalar. If there are in fact several unknown parameters, then their number must be reduced to 1 by the imposition of invariance requirements. The X_i

are assumed to be independent, and the alternative decisions are:

$$
\begin{array}{ll}
D_0 : \theta_i = \theta & i = 1 \text{ to } n \\
D_i : \theta_i = \theta + \Delta & \Delta > 0 \\
\theta_j = \theta & j \neq i
\end{array}
$$

Under mild assumptions, Karlin and Truax show that any symmetric Bayesian procedure to the slippage problem is of the form:

Conclude D_0 if $h(X_1 \ldots X_n) \in R_0$

Conclude D_i if $h(X_1 \ldots . X_n) \notin R_0$ and $X_i > X_j$ for all $j \neq i$ (2.1)

where h is a function symmetric in $X_1 \ldots . X_n$

An assumption implicit in using a test that rejects the largest order statistic is that the alternative hypothesis leads to outliers on the right of the sample. This is formalized by Karlin and Truax in their assumption that f satisfies for all k

$$\left\{ \prod_{\substack{j=1 \\ j \neq i}}^{n} f(X_j|\theta) \right\} f(X_i|\theta + \Delta) \geqslant \left\{ \prod_{\substack{j=1 \\ j \neq k}}^{n} f(X_j|\theta) \right\} f(X_k|\theta + \Delta)$$

This general theory applies to the case of a single outlier, specified *a priori* to be on the right if it occurs at all. Of course the theory may be rephrased trivially to cover the case of an outlier known to be on the left.

As special cases, Karlin and Truax deduce the following results:

(i) If $X_i \sim N(\theta_i, \sigma^2)$ where σ is known, then the optimal test leads to D_i if $X_i = X_{(n)}$ and $(X_{(n)} - \bar{X}) > C\sigma$ where $X_{(1)}, X_{(2)}, \ldots, X_{(n)}$ denote the other statistics of the sample ranked from smallest to largest.

(ii) If σ is unknown, then there are two unknown parameters, and one must be eliminated by invariance. It is intuitively reasonable to study only procedures which are invariant under linear transformation of the X_i, and such procedures must depend on the X_i only through the maximal invariants $U_1 \ldots U_n$ where

$$U_i = (X_i - \bar{X})/s$$

From this it follows that the optimal decision is D_i if $X_i = X_{(n)}$ and $(X_{(n)} - \bar{X}) > Cs$

In fact, Karlin and Truax's solution to this problem is slightly more general than is indicated above. They assume that we have available

nk observations from a balanced one-way analysis of variance,

$$Y_{ij} \quad j=1 \text{ to } k, \quad i=1 \text{ to } n$$
$$Y_{ij} \sim \mathrm{N}[\theta_i, k\sigma^2]$$
$$X_i = Y_{i.} \sim \mathrm{N}[\theta_i, \sigma^2]$$

and

$$s^2 = \sum_i \sum_j (Y_{ij} - Y_{i.})^2/k + \sum_i (X_i - \bar{X})^2$$

This formulation is identical to that of Paulson (1952) who found by direct methods that the test procedure set out above is optimal. A related formulation is that of Kudô (1956) who merely assumes that some information on σ, external to the sample, is available in the form of an independent $\sigma^2\chi^2$ variable. For this formulation, essentially the same optimal statistic arises. Thus this result of Karlin and Truax has both the Paulson and the Kudô results as special cases.

Two every general results also follow when θ is a location parameter $f(x|\theta) = f(x-\theta)$. If θ admits a sufficient maximum likelihood estimator (MLE) $\hat{\theta}$, then the optimal procedure concludes D_i if

$$X_i = X_{(n)} \quad \text{and} \quad (X_{(n)} - \hat{\theta}) > C \tag{2.2}$$

Similarly, if θ is a scale parameter $f(x|\theta) = f(x/\theta)$ and has a sufficient maximum likelihood estimator $\hat{\theta}$, then the optimal procedure concludes D_i if

$$X_i = X_{(n)} \quad \text{and} \quad X_{(n)}/\hat{\theta} > C \tag{2.3}$$

This generalizes the problem discussed in Truax (1953) of testing whether a set of χ^2 variates has slipped to the right: specifically,

$$X_i/\theta_i \simeq \chi^2_v$$

(note the common number of degrees of freedom). To test whether one θ_i has slipped to the right, the optimal procedure uses Cochran's (1941) test: decide D_i if

$$X_i = X_{(n)} \quad \text{and} \quad X_{(n)} \Big/ \sum_{i=1}^{n} X_i > C \tag{2.4}$$

Finally, if θ and σ are a location–scale pair of parameters

$$f(x|\theta, \sigma) = f\{(x-\theta)/\sigma\}/\sigma$$

and θ and σ have jointly sufficient maximum likelihood estimators, then the optimal decision is D_i if

$$X_i = X_{(n)} \quad \text{and} \quad (X_{(n)} - \hat{\theta})/\hat{\sigma} > C \tag{2.5}$$

From this, the result above for the normal distribution may be deduced. This was shown by Truax to be optimal, a result deduced as a special case of Karlin and Truax's theory.

Unfortunately, despite the general nature of these results, their specific application is not as broad as it seems. The fact that the general results on location and on scale parameters only hold for a parameter having a sufficient statistic reduces the possible distributions to two types:

(i) those where θ is a range parameter – for example, the shifted exponential

$$f(x|\theta) = \exp[-(x-\theta)] \qquad x > \theta$$

(ii) the regular exponential family

$$f(x|\theta) = \exp[a(x)b(\theta) + c(x) + d(\theta)]$$

Some interesting results derive from the application of the general theory to distributions of type (i). For example, in the shifted exponential

$$f(x|\theta) = \exp[-(x-\theta)] \qquad x > \theta$$

the sufficient MLE for θ is $X_{(1)}$. Thus the optimal test for an outlier rejects on large values of $X_{(n)} - X_{(1)}$ – a statistic proposed by Basu (1965).

Similarly, if $f(x|\theta, \sigma) = \exp[-x-\theta)/\sigma]/\sigma$ then sufficient MLEs for θ and σ are

$$\hat{\theta} = X_{(1)}$$

$$\hat{\sigma} = \sum_{2}^{n} (X_{(i)} - X_{(1)})/n$$

and an optimal test statistic is

$$(X_{(n)} - \hat{\theta})/\hat{\sigma}$$

The other class of distributions to which the theory applies, however, is rather a disappointment. Apart from the normal and gamma distributions, the regular exponential family turns out to have few interesting members having location or scale parameters as defined above.

The results of Karlin and Truax (1960) are specific to a distribution with only one unknown parameter. Nuisance parameters are dealt with the imposing invariance requirements. While it is no doubt true that, in most applications, sensible suitable invariance requirements are apparent, it nevertheless seems appropriate to seek some theory which is not dependent on such a specification. The first section of Hawkins (1976) suggests as a method of dealing with nuisance parameters that the optimal outlier statistic be found from the (parameter-free) conditional distribution of the X given the possibly vector-valued sufficient statistic for any unknown parameters. While this avoids the arbitrariness of the invariance requirement, the approach is still specific to distributions in the general exponential family. In the widely studied normal and gamma families, this procedure yields identical procedures to those mentioned above.

Several other optimality results have been proved for specific underlying distributions. Murphy (1951), in one of the few theoretical results proved for multiple outlier tests, considers the model

$$X_{j(i)} \sim \mathrm{N}[\theta + a_i \delta, \sigma^2] \tag{2.6}$$

where the a_i are known, $a_1 \leqslant a_2 \leqslant \ldots \leqslant a_n$ and $j(i)$ is some unknown permutation of the integers $1, 2, \ldots, n$. He shows that an optimal test in the sense of maximizing the power subject to reasonable invariance requirements uses as test statistic $\Sigma_i a_i (X_{(i)} - \bar{X})/s$ where $s^2 = \Sigma_i (X_i - \bar{X})^2$. As a special case of this result, if $a_1 = a_2 = \ldots = a_{n-1} = 0$, $a_n = 1$, we have the optimality of $(X_{(n)} - \bar{X})/s$. Another special case is the optimality for $a_1 = -1$, $a_2 = a_3 = \ldots = a_{n-1} = 0$, $a_n = 1$ of the studentized range $(X_{(n)} - X_{(1)})/s$. This latter result was also proved by Ramachandran and Khatri (1957).

Probably the most widely used of the criteria flowing from Murphy's work is the statistic $(X_{(n)} + X_{(n-1)} - 2\bar{X})/s$, used for testing a pair of possible outliers on the same side of the sample.

If the a_i are not known *a priori*, Murphy shows that the generalized likelihood ratio test for k outliers is based on the ratio $s_{1,2,\ldots,k}/s$, where $s_{1,2,\ldots,k}$ is the standard deviation obtained when the k most extreme observations are deleted from the sample. This statistic generalizes one for the case of $k = 2$ whose distribution is studied in Grubbs (1969) and has more recently been proposed again by Tietjen and Moore (1972).

This generalized likelihood ratio approach deserves a little further comment since it is very widely applicable, but has seldom been used to derive good outlier tests. The procedure consists of maximizing the likelihood in two stages: at the first, assuming a particular permutation $j(i)$, one maximizes the likelihood analytically over any free parameters. At the second stage, one chooses as the outliers that permutation $j(i)$ for which this maximized likelihood is a maximum. It is a matter of interest that, in common with other applications of generalized likelihood ratio tests, this method does not produce a test statistic with any guaranteed optimality properties. Nevertheless, in actual practice it does quite commonly yield those same outlier statistics shown to be optimal by direct methods.

A different criterion is used by Ferguson (1961a). Noting the general non-existence of uniformly most powerful outlier tests unless additional constraints such as invariance are imposed, he addresses the problem by looking for locally most powerful tests in the neighbourhood of the null hypothesis. The null hypothesis is

$$H_0 : X_i \sim \mathrm{N}(\theta, \sigma^2) \qquad i = 1 \text{ to } n$$

Two possible alternative hypotheses are

$$H_1 : \text{Several } X_i \sim \mathrm{N}(\theta + \delta_i\sigma, \sigma^2) \qquad \delta_i > 0$$
$$\text{the remaining } X_i \sim \mathrm{N}(\theta, \sigma^2)$$

and

$$H_2 : \text{Several } X_i \sim N(\theta, \lambda_i^2\sigma^2) \qquad \lambda_i > 1$$
$$\text{the remaining } X_i \sim \mathrm{N}(\theta, \sigma^2)$$

Ferguson shows that for any number of outliers up to $n/2$, the power against H_1 in the neighbourhood of $\boldsymbol{\delta} = \mathbf{0}$ is maximized by using the critical region $\sqrt{b_1} > C$ where $\sqrt{b_1}$ is the sample moment coefficient of skewness,

$$\sqrt{b_1} = \sum_1^n \{(X_i - \bar{X})/s\}^3/n$$

of course if $\boldsymbol{\delta}$ is known to be negative, then one rejects if $-\sqrt{b_1} > C$.

Notice that this test statistic, unlike the β_1-optimal tests introduced earlier in this chapter, does not test whether a specific datum is an outlier. It merely indicates that one or more outliers are present. To turn it into a rule for identifying outliers, it must be supplemented by another rule for rejection – one possible heuristic would be to reject extreme observations until $\sqrt{b_1}$ becomes non-significant.

If the alternative hypothesis H_2 holds, then the contaminants can occur on either the left or the right of the sample. For this situation, and for the situation in which the sign of $\boldsymbol{\delta}$ is unknown, the locally best test rejects if the sample moment of kurtosis

$$b_2 = \sum_1^n \{(X_i - \bar{X})/s\}^4/n \geqslant C$$

This is true provided the proportion of outliers does not exceed a maximum dependent on n, but never less than 21 per cent.

Unlike the Paulson (1952) result, these procedures are locally optimal regardless of the number of outliers, so long as this is not too large. However, the practical use of the statistics is hampered by the need for an auxiliary rule if any actual classification of outliers is required. Note also that as $X_{(n)}$ tends to ∞, both $\sqrt{b_1}$ and b_2 tend to become monotonic functions of $(X_{(n)} - \bar{X})/s$ alone, suggesting some sort of asymptotic equivalence for a single outlier far out on the right of the remainder of the sample.

The optimality proved by Ferguson is local (near $\boldsymbol{\delta} = \mathbf{0}$ or $\boldsymbol{\lambda} = \mathbf{1}$). He also showed by simulation that if the values of $\boldsymbol{\delta}$ or $\boldsymbol{\lambda}$ are not close to those specified by H_0 then $(X_{(n)} - \bar{X})/s$ is more powerful than $\sqrt{b_1}$, and $\max|X_i - \bar{X}|/s$ more powerful than b_2 for detecting a single outlier; the difference in powers is not large, however.

The power of the skewness and kurtosis statistics against multiple outliers, however, provides a strong motivation for using them to screen data sets routinely, preparatory to any formal analysis aimed at identifying outliers.

2.3 Use of 'departure from model' statistics

The sample skewness and kurtosis are usually considered as test statistics useful for testing whether a sample is normal, and indeed the presence of outliers is one way in which the distribution could depart from normality. This suggests the possibility of using any goodness-of-fit statistic as the basis for an outlier test. To put this in a general framework, let $F_n(x)$ be the sample cumulative distribution function of $X_1 \ldots X_n$, and $F_0(x)$ the supposed true distribution function. Any measure of fit of $F_n(x)$ to $F_0(x)$ may be used. One example in the case of the normal distribution is the Shapiro–Wilk statistic W, which is essentially the same as

$$\int_{-\infty}^{\infty} [F_n(x) - F_0(x)]^2 \, dF_0(x)$$

This statistic is shown by Tiku (1975) not to be particularly powerful. A better statistic would be one that emphasizes the quality of fit in the tails. Such a statistic is the Anderson–Darling statistic

$$A_n^2 = n \int_{-\infty}^{\infty} \frac{[F_n(x) - F_0(x)]^2}{F_0(x)[1 - F_0(x)]} \mathrm{d}F_0(x)$$

A discussion of the null distribution of A_n^2 is given in Anderson and Darling (1954). The possibility of using A_n^2 as an outlier test does not seem to have been investigated. However, the idea is appealing since the statistic is completely general, and may be used for any supposed underlying distribution $F_0(x)$.

A standard representation of A_n^2 is

$$A_n^2 = -n - n^{-1} \sum_{j=1}^{n} (2j - 1)[\log F_0(X_{(j)}) + \log\{1 - F_0(X_{(n-j+1)})\}] \qquad (2.7)$$

which illustrates how sensitive it is to the extreme order statistics occurring in regions of low density under the null hypothesis. For example, if $F_0(.)$ is the N(0, 1) density then A_n^2 varies asymptotically as $X_{(n)} \to \infty$ as $\frac{1}{2}n^{-1}X_{(n)}^2$, and is thus fairly sensitive to the possibility of $X_{(n)}$ being an outlier. If, however, several outliers are present, then it becomes even more sensitive, and hence competitive.

Tiku (1975) proposes a general family of statistics which, while aimed at testing for general departures from model, are likely to be more effective for detecting outliers and heavy tails than other model departures. The statistics are applicable to any distribution having a scale parameter σ. The test statistic is the ratio of two estimators $\hat{\sigma}$ and $\hat{\sigma}_c$ where $\hat{\sigma}$ is an estimator of σ based on all order statistics, while $\hat{\sigma}_c$ is based only on those central order statistics believed *a priori* to be reliable.

On a less formal plane, the use of graphic methods such as, for example, a stem-and-leaf plot, or probability plot of residuals from a regression, may provide a useful indication of whether outliers are present, or whether the underlying distribution was merely heavier-tailed than had been supposed. Such methods may be formalized into a large-sample non-parametric gap test, which will be set out in Chapter 6.

2.4 Inadequacies of current optimality theory

Despite the body of theoretical optimality results set out earlier, there are several questions to which a complete answer has not yet been given. The major one of these is how to proceed in a situation with possible multiple outliers whose number is unknown. There are on the one hand the general results of Ferguson (1961a) showing that $\sqrt{b_1}$ and b_2 are good statistics for any reasonable number of outliers. These statistics, and other goodness-of-fit statistics, do not, however, identify the outlier(s) explicitly and so must be supplemented with some heuristic which enables them to locate outliers. On the other hand, the tests proposed by Murphy (1951) require prior knowledge of the number of outliers, while those covered by Karlin and Truax's (1960) general theory are specific to the case of a single outlier.

Another problem is that the theory generally assumes $X_1, \ldots, X_n$ to be independent, identically distributed random variables. Several of the results still hold if the X_i are merely exchangeable, but there is no entirely satisfactory theory of optimal tests if the X_i are not exchangeable. Thus two extremely interesting problems have no known optimal solution:

(i) Given $X_i \sim \sigma_i^2 \chi_{v_i}^2$, $i = 1$ to n with not all v_i equal, test whether one σ_i differs from the others. This is just the problem of deciding in an analysis of variance whether one of the populations has slipped in variance. As already noted, if the v_i are equal, Cochran's (1941) statistic $X_{(n)}/\Sigma X_i$ is optimal, but for unequal v_i the optimal statistic is not known.

(ii) Given a general linear hypothesis

$$\mathbf{Y} = \mathbf{X}\boldsymbol{\beta} + \mathbf{e} \qquad \text{where} \qquad \mathbf{e} \sim \mathrm{N}(\mathbf{0}, \sigma^2\mathbf{I})$$

the residual vector

$$\hat{\mathbf{e}} = \mathbf{Y} - \mathbf{X}\boldsymbol{\beta}$$

has elements which are interdependent and, in general, not exchangeable. In the non-exchangeable case, optimal tests for locating outliers amongst the e_i are not known.

Hawkins (1976) provides some theory showing how to find optimal test statistics for outliers amongst non-exchangeable variables having distributions in the general exponential family; however, an assumption is made implicitly that an error in the ith variable will tend to displace it by an amount independent of i – a situation that is by no means always reasonable to assume.

2.5 Discrete random variables

The results given earlier depend in quite a basic way on the continuity of X. This is required for setting up the pivotal quantities needed to yield similar tests in the face of one or more unknown nuisance parameters. If X is discrete, then the necessary pivotal quantities do not exist and the theory breaks down. A solution seems possible in two cases. When all parameters are known (as, for example, with the hypergeometric distribution) then the problem of unknown or nuisance parameters does not arise and the optimal procedure is clearly to test $X_{(n)}$ against its (completely known) distribution.

The other soluble case arises if the distribution of X has a single unknown parameter, but is a member of the generalized power series distribution (GPSD) family. This family includes the Poisson, binomial and negative binomial distributions. In this case, a sufficient statistic T for the unknown parameter exists, and the optimal test is obtained by using the conditional distribution of $X_{(n)}$ given T, which is known completely.

2.6 Relationship with tolerance regions

Another theoretical avenue that seems not to have been explored extensively is the relationship between outlier test statistics and tolerance regions. The general theory of tolerance regions is set out in Guttman (1970), to which source the interested reader is referred. For present purposes, the interesting problem in the theory of tolerance regions is: we are given $X_1, \ldots, X_n, Y$ which are $n+1$ independently and identically distributed (i.i.d.) random variables from a density $f(x|\theta)$. The form of f is known, but the (possibly vector-valued) parameter θ is not. A γ expectation tolerance region is defined as a region $S(X_1, \ldots, X_n)$ satisfying

$$E[\Pr\{Y \in S(X_1, \ldots, X_n)\}] = \gamma$$

regardless of θ.

Paulson (1943) defines a tolerance region as optimal if it is the smallest amongst all tolerance regions of the same expectation, and shows how optimal tolerance regions may be constructed from standard optimal confidence intervals.

Thus an intuitively sensible procedure for outlier detection would be: select a suitable γ; for each i, use $X_1, X_2, \ldots, X_{i-1}, X_{i+1}, \ldots, X_n$ to set up a γ expectation tolerance region S_i for X_i. If any X_i is not in S_i then conclude that at least one outlier is present. Identify the outlier by increasing γ until only one X_i lies outside its tolerance region. This X_i is then identified as the outlier.

It is extremely interesting to find that this procedure yields the same test as the β_1 optimal test in all the situations discussed in this book. It also yields a result for X_i that are not identically distributed, as will be illustrated in Chapter 7.

Hawkins (1976) gives a motivation for believing that this correspondence between β_1 optimal outlier tests and optimal tolerance regions is not fortuitous, but the consequence of a unifying correspondence between the two approaches.

2.7 Null hypothesis distributions

Broadly speaking, we can distinguish between two classes of outlier statistics – those which test for the presence of outliers but do not identify particular observations as outliers, and those in which identification of the outlier is implicit. The former class includes such statistics as $\sqrt{b_1}$, b_2 and A_n^2. The distribution of these statistics may be complicated but does not in general involve order statistic distributions. The statistics in the latter class are typically of the form

$$T = \max_{1 \leq i \leq n} h_i(X_i, \mathbf{U})$$

where $\mathbf{U}$ is a statistic based on the entire sample. If the distribution of $h_i(X_i, \mathbf{U})$ for an arbitrary X_i is well understood, then that of T may be deduced by means of the identity

$$\Pr[T > t] = \Pr\left[\bigcup_{i=1}^{n} h_i(X_i, \mathbf{U}) > t\right]$$

$$= \Pr\left[\bigcup_{1}^{n} E_i\right] \qquad \text{say}$$

where E_i is the event $h_i(X_i, \mathbf{U}) > t$.

$$\Pr[\cup E_i] = \sum_i \Pr[E_i] - \sum_i \sum_j \Pr[E_i \cap E_j] + \ldots$$

$$- (-1)^n \Pr[E_1 \cap E_2 \cap \ldots \cap E_n]$$

An inequality due to Boole but known generally as Bonferroni's inequality states that any two successive partial sums on the right-hand side of this expansion bracket the probability on the left. In particular,

$$\sum_i \Pr[E_i] - \sum_i \sum_j \Pr[E_i \cap E_j] \leqslant \Pr[\cup E_i] \leqslant \sum_i \Pr[E_i] \qquad (2.8)$$

and choosing t so that $\Sigma \Pr[E_i] = \alpha$ yields a test on T whose size does not exceed α. For many outlier tests, it can be shown that this conservative approximation is extremely good. It is frequently exact and when not exact typically yields an overstatement of the size of the test of not more than $\frac{1}{2}\alpha^2$ (see for example Doornbos, 1959).

In the case of i.i.d. variables, if we let $g(.)$ denote the density of $h(X_i, \mathrm{U})$; $f_n(.)$ that of T, and $F_n(.)$ the cumulative distribution function of T, then Hawkins (1976) shows that under certain, often-satisfied conditions, the exact distribution of T may be found recursively in the form

$$f_n(t) = ng(t)F_{n-1}\{H_n(t)\} \tag{2.9}$$

Particular cases of this recursion will be mentioned again in Chapters 3 and 4. The function $H_n(t)$ is determined algebraically from the form of the test statistic.

2.8 Relations between measures of performance

Earlier, three measures of the performance of an outlier test were defined for an alternative hypothesis in which one contaminant is present (for convenience of notation, we assume it is X_1). The measures were

$$\beta_1 = \Pr[\text{reject } H_0 \text{ and identify } X_1 \text{ as the outlier}]$$
$$\beta_2 = \Pr[\text{reject } H_0]$$
$$\beta_3 = \Pr[\text{reject } H_0 \mid X_1 \text{ is the value tested}]$$

In the common situation the test is of the form $T = \max h_i(X_i, \mathrm{U})$; reject H_0 if $T > k$ and classify as an outlier that X_i having the largest h_i. In this case, the two remaining measures of performance defined by David and Paulson (1965) are

$$\beta_4 = \Pr[h_1 > k]$$
$$\beta_5 = \Pr[h_1 > k \text{ and } h_j < k \text{ for all } j > 1] \quad \text{where } h_i = h_i(X_i, \mathrm{U})$$

We will say no more about β_5, but β_4 is very useful since it is usually relatively easy to evaluate, its distribution not depending on order statistic distributions. Now it follows at once from these definitions that $\beta_2 \geqslant \beta_4 \geqslant \beta_1$. To derive useful approximations suppose also that:

(i) under H_0, if k is the α fractile of T,

$$\Pr[(h_i > k) \quad \text{and} \quad (h_j > k)] = o(\alpha)$$

(ii) under H_1, if $j > 1$, $\Pr[h_j > k | H_1] < \Pr[h_j > k | H_0]$

These assumptions are by no means true universally. Suppose for example, that $n = 2$ and under H_0, $X_i \sim N(\theta, 1)$ while under H_1, the mean of X_1 has slipped in an unknown direction. Then $h_i = |X_i - \bar{X}|$ is the optimal statistic, but $h_1 = h_2$. Thus both assumptions fail.

If the assumptions do hold, then the following results may be deduced:

$$\begin{aligned}\beta_4 - \beta_1 &= \Pr[h_1 > k \quad \text{and} \quad h_j > h_1 \text{ for some } j > 1] \\ &< \Pr[h_1 > k \quad \text{and} \quad h_j > k \text{ for some } j > 1] \\ &< \Pr[h_j > k \quad \text{for some} \quad j > 1] \\ &< (n-1)\Pr[h_1 > k] \\ &< (n-1)(\alpha/n + o(\alpha)) \qquad \text{by (i) above} \\ &< \alpha + o(\alpha) \end{aligned} \tag{2.10}$$

Next,

$$\begin{aligned}\beta_2 &= \Pr[\cup h_j > k] \\ &= \Pr[h_1 > k \cup (h_j > k \text{ for some } j > 1)] \\ &< \beta_4 + (n-1)\Pr[h_j > k] \\ &< \beta_4 + (n-1)(\alpha/n + o(\alpha)) \\ &< \beta_4 + \alpha + o(\alpha) \end{aligned} \tag{2.11}$$

Thus if assumptions (i) and (ii) hold we have $\beta_4 + \alpha + o(\alpha) > \beta_2 > \beta_4 > \beta_1 > \beta_1 > \beta_4 - \alpha$, and so for practical purposes, the three measures are equivalent. In any particular case, it remains to verify the two assumptions. In many cases, much tighter inequalities can be proved between β_1, β_2 and β_4.

As noted already, the measure β_3 has not found wide acceptance; apart from the trite inequality $\beta_3 > \beta_1$, little can be said in full generality about its relationship with the other measures of performance. In the usual cases, hower, β_3 will tend to 1 as the contaminating distribution becomes more aberrant.

CHAPTER 3

A single outlier in normal samples

In view of the central position which the normal distribution occupies in parametric statistics, it is not surprising to find that the question of outliers from normal samples has received both the earliest and the most concentrated study in outlier theory. This chapter will discuss a very particular class of problems – those in which the data are assumed, under the null hypothesis, to be both normal and independent, and identically distributed. A later chapter will relax this assumption of identical distribution by assuming that the data are generated by a homoscedastic general linear model. It will also be assumed that, if the null hypothesis is false, at most a single outlier is present. This assumption will also be relaxed in a later chapter.

We assume then $X_1, \ldots, X_n$ all independently and identically distributed as $N(\xi, \sigma^2)$. We will also assume the existence of some additional information about σ^2 in the form of U, an independent $\sigma^2\chi^2_\nu$ variable.

In some cases, no external information on σ^2 is available – if this occurs, we shall regard U and ν as zero in the general theory. Furthermore, some statistics disregard the availability of U, and for these tests, the same convention will be applied.

In other cases, σ may be known *a priori*. One method of handling this situation is setting $U = \nu\sigma^2$ in the general theory, and letting ν tend to infinity.

The major factors influencing the outlier test statistics to be used are:

(i) whether or not ξ and σ are known *a priori*;
(ii) whether or not it is known on which side of the sample any contaminants are likely to occur.

3.1 Notation

Let $$\bar{X} = \sum_1^n X_i/n \quad s^2 = \sum_1^n (X_i - \bar{X})^2$$

Analogously, let $\bar{X}_{i,j,\ldots,k}$ and $s^2_{i,j,\ldots,k}$ denote the mean and sum of squared deviations about the mean of the sample obtained by deleting $X_i, X_j, \ldots, X_k$ from the original sample.
Define $S = s^2 + U$; $S_{i,j,\ldots,k} = s^2_{i,j,\ldots,k} + U$
Let $X_{(1)} \leqslant X_{(2)} \leqslant \ldots \leqslant X_{(n)}$ denote the n order statistics of the sample.

3.2 Optimal statistics

The early writings on outliers assumed for the most part that both ξ and σ were known, and recommended rejection of any X_i for which $|X_i - \xi|/\sigma$ was unduly large. In actual practice, of course, ξ and σ were to be replaced by sample estimates, but the effect of this was not understood. Of course, with both parameters known, the problem of the distribution theory may be reduced to that of the distribution of the extreme order statistics from a N(0, 1) distribution.

Two moves in the direction of greater realism came in the papers of McKay (1935) and of Thompson (1935). McKay, considering the cumulants of the statistic $(X_{(n)} - \bar{X})/\sigma$, was able to derive the following result.

Let $f_n(.)$ denote the density of $(X_{(n)} - \bar{X})/\sigma$, and $F_n(.)$ the corresponding cumulative distribution function. Then

$$f_n(x) = n\{n/2\pi(n-1)\}^{1/2} \exp[-nx^2/2(n-1)] \cdot F_{n-1}\{nx/(n-1)\} \tag{3.1}$$

a special case of the general recursion (2.9).

Using the starting condition

$$F_1(x) = \begin{cases} 0 & x < 0 \\ 1 & x \geqslant 0 \end{cases}$$

the densities $f_n(x)$ and cumulative distribution functions $F_n(x)$ could then be derived for all n and x.

This distributional result was soon derived independently by several other workers. Godwin (1945), Nair (1948) and Grubbs (1950) derived the same recursion in the following different way. Note first that the joint distribution of the n order statistics from a N(0, 1) sample is

$$f(x_{(1)}, \ldots, x_{(n)}) = n!/(2\pi)^{n/2} \exp\left[-\frac{1}{2}\sum_1^n X_i^2\right] \tag{3.2}$$

Now make the Helmert orthogonal transformation

$$\begin{aligned}
Y_1 &= -(X_{(1)} - X_{(2)})/2^{1/2} \\
Y_2 &= -(X_{(1)} + X_{(2)} - 2X_{(3)})/6^{1/2} \\
&\;\;\vdots \\
Y_{n-1} &= -(X_{(1)} + X_{(2)} + \ldots + X_{(n-1)} - (n-1)X_{(n)})/\{n(n-1)\}^{1/2} \\
Y_n &= (X_{(1)} + X_{(2)} + \ldots + X_{(n)})/n^{1/2}
\end{aligned} \tag{3.3}$$

which, by orthogonality, gives the joint distribution of the Y_i as

$$f(Y_1, \ldots, Y_n) = n!/(2\pi)^{n/2} \exp\left[-\frac{1}{2}\sum_1^n Y_i^2\right] \tag{3.4}$$

The original ranking constraint $X_{(i)} \geqslant X_{(i-1)}$ becomes $Y_r\sqrt{(r+1)} > Y_{r-1}\sqrt{(r-1)}$. On integrating out the joint density, McKay's recursion results.

Grubbs (1950) programmed the recursion on ENIAC, and produced an extensive and still useful table of $F_n(x)$ for $n = 2(1)25$ and $x = 0\ (.05)\ 4.90$.

David (1956) followed up the suggestion by McKay that the Bonferroni inequality be used as an approximation to the fractiles of $(X_{(n)} - \bar{X})/\sigma$. He noted the high accuracy of the approximation, and also discussed the power of the resulting test.

The other important result of 1935 was the derivation by Thompson of the distribution of an arbitrary studentized residual $(X_i - \bar{X})/\{S^2/(n-1)\}^{1/2}$. As the distribution will be of importance later on, it is convenient to derive the distribution in somewhat greater generality than was used by Thompson. It is easily verified that

$$(X_i - \bar{X}) = (n-1)(X_i - \bar{X}_i)/n$$

and (3.5)

$$s^2 = s_i^2 + (n-1)(X_i - \bar{X}_i)^2/n$$

Thus adding U to both sides,

$$S = S_i + (n-1)(X_i - \bar{X}_i)^2/n$$

Now clearly $(X_i - \bar{X}_i)\ \{(n-1)/n\}^{1/2} \sim \mathrm{N}(0, \sigma^2)$ and $S_i \sim \sigma^2\chi^2_{n+\nu-2}$

are independent. Thus

$$t_i = \{(n-1)(n+v-2)/n\}^{1/2}(X_i - \bar{X}_i)/\sqrt{S_i}$$

follows a t-distribution with $n+v-2$ degrees of freedom and density

$$f(t) = \frac{\Gamma[\frac{1}{2}(n+v-1)]}{\Gamma[\frac{1}{2}(n+v-2)][\pi(n+v-2)]^{1/2}} \frac{1}{[1+t^2/(n+v-2)]^{(n+v-1)/2}}$$

Letting $r_i = (X_i - \bar{X}_i)/\sqrt{S_i}$, this has density

$$f(r) = \frac{\Gamma[\frac{1}{2}(n+v-1)](n-1)^{1/2}}{\Gamma[\frac{1}{2}(n+v-2)](\pi n)^{1/2}} \frac{1}{[1+(n-1)r^2/n](n+v-1)/2}$$

and hence $b_i = (X_i - \bar{X})/\sqrt{S}$ has density

$$h(b) = \frac{\Gamma[\frac{1}{2}(n+v-1)]n^{1/2}}{\Gamma[\frac{1}{2}(n+v-2)][\pi(n-1)]^{1/2}}[1 - nb^2/(n-1)]^{(n+v-4)/2} \quad (3.6)$$

a result which follows from the identity

$$b_i = (n-1)r_i/[n\{n+(n-1)r_i^2\}]^{1/2} \quad (3.7)$$

Now Thompson's result follows when one sets $v = 0$ and makes the further change of variable to $(n-2)^{1/2} b_i$.

From this distribution, fractiles of b_i could be derived quite easily. The effort involved is, however, somewhat wasted, since these fractiles could be deduced from those of t_i, which follows the familiar and well-tabulated t-distribution, or from the distribution of the correlation coefficient since (as is easily seen and justified), $b_i\{(n-1)/n\}^{1/2}$ has the distribution of a correlation coefficient in a bivariate normal sample.

Thompson's suggestion was that any b_i be rejected which exceeded the ρ/n fractile. He envisaged that this procedure, by rejecting an average of ρ good observations per sample, could be made quite effective against outliers. Tables for $\rho = 0.2$, 0.1 and 0.05 and a number of n values were included in his paper.

The next major development in this direction came in the paper of Pearson and Chandra Sekar (1936). Following on from the work of Thompson, they made the additional observation that for any $i \neq j$, the events $b_i > c_1$ and $b_j > c_1$ are mutually exclusive, as are $|b_i| > c$, and $|b_j| > c$, where $c = 0.5^{1/2}$ and $c_1 = \{(n-2)/2n\}^{1/2}$. While the main conclusion drawn from this was that the b_i are inefficient in locating multiple outliers (because of the masking effect), the finding also permitted an exact calculation of some fractiles of the test statistic

$$B = \max b_i$$

These fractiles come about from the fact that

$$\begin{aligned}\Pr[B > B_0] &= \Pr[\text{at least one } b_i > B_0] \\ &= n\Pr[b_i > B_0] \qquad (3.8)\end{aligned}$$

provided $B_0 > c_1$, for then the second and subsequent terms in the Boole expansion, all of which involve joint probabilities of more than one b_i exceeding B_0, become zero. Thus, if B_0 is the α/n fractile of the distribution of an arbitrary b_i, and if B_0 exceeds c_1, then B_0 is the α fractile of B.

All these results were shown by Pearson and Chandra Sekar for the case $U = v = 0$. However, it is a simple matter to adapt their results to the more general case. The limits c and c_1 are unaltered, and one uses the α/n fractile of $h(b)$.

While Pearson and Chandra Sekar did not say so explicitly, their method also yields some fractiles of the statistic

$$B^* = \max|b_i|$$

Here, provided $B_0^* > c$,

$$\Pr[B^* > B_0^*] = 2n\Pr[b_i > B_0^*]$$

and so the $\alpha/2n$ fractile of $h(b)$ provides the α fractile of B^*.

For $U = v = 0$, this method provides exact fractiles for a number of small to moderate values of n, for example, the 5 per cent fractile of B for $n < 15$ and of B^* for $n < 17$.

The now common use of the Bonferroni inequality seems not to have been applied to the statistic B until 1959, when Doornbos studied the statistic

$$T = \max t_i$$

Apparently unaware of the fact that T and B are monotonic functions of one another, Doornbos showed that, for cases of practical interest, for arbitrary i and j

$$\Pr[t_i > T_0 \cap t_j > T_0] < \{\Pr(t_i > T_0)\}^2$$

From this, if T_0 is the α/n fractile of the t-distribution, the first two Bonferroni inequalitites show that

$$\alpha - \tfrac{1}{2}\alpha^2 < \Pr[T > T_0] \leqslant \alpha$$

a result showing that, even when the Pearson–Chandra Sekar fractile is not exact, it is still an excellent, and slightly conservative approximation to the true fractile.

Finally, Quesenberry and David (1961) introduced the general statistics B and B^* with v possibly nonzero. They found the joint distribution of an arbitrary pair b_i, b_j to be

$$f(b_i, b_j) = (n/\{n-2\})^{1/2}(n+v-3)D^{n+v-5}/(2\pi) \qquad D^2 > 0 \quad (3.9)$$

where

$$D^2 = 1 - \{(n-1)(b_i^2 + b_j^2) + 2b_i b_j\}/(n-2)$$

Using the approximation

$$\Pr[B > B_0] \approx n\Pr[b_i > B_0] - \tfrac{1}{2}n(n-1)\Pr[(b_i > B_0)\cap(b_j > B_0)]$$

they derived even more accurate (though not conservative) approximate fractiles for B and, using an exactly analogous approximation, for B^*.

In the meantime, Grubbs (1950) derived an exact expression for the density of B in the case $U = v = 0$. Starting off from the Helmert transformation which, as we have already observed, leads to the density of $(X_{(n)} - \bar{X})/\sigma$, he made a further transformation to spherical polar co-ordinates, and deduced from this the distribution of

$$D = s_{(n)}^2/s^2$$

Since $D = 1 - nB^2/(n-1)$, and B is known to be positive, Grubbs's fractiles for the D-distribution yielded at once fractiles for B. Grubbs's expression for the cumulative distribution function of D was in terms of a rather complicated multiple integral.

Borenius (1959, 1966) studied the distribution of B with $U = v = 0$ and arrived at what appears to be the same multiple integral as Grubbs's, though with different variables of integration.

Hawkins (1969a) extended these results to the case v possibly nonzero, and obtained the following simpler expression. Letting $G_{n,v}(.)$ denote the cumulative distribution function of B based on sample size n and v external degrees of freedom, as a further special case of the recursion (2.9)

$$G_{n,v}(x) = \begin{cases} \int_0^{g(x)} nf(r)G_{n-1,v}(r)\,dr & x < \{(n-1)/n\}^{1/2} \\ 1 & x \geqslant \{(n-1)/n\}^{1/2} \end{cases} \qquad (3.10)$$

where $g(x) = nx/[(n-1)\{1 - nx^2/(n-1)\}^{1/2}]$

and $f(r)$ is the density of r, shown earlier to be

$$f(r) = \frac{\Gamma[\frac{1}{2}(n+v-1)](n-1)^{1/2}}{\Gamma[\frac{1}{2}(n+v-2)](\pi n)^{1/2}} \frac{1}{[1+(n-1)/r^2/n]^{(n+v-1)/2}}$$

From this recursion, the distribution function of B can be computed quite easily by numerical integration and any desired fractiles computed.

The distribution of B^* is considerably more difficult to find. Hawkins and Perold (1977) have derived a recursive expression for the joint distribution of $(X_{(1)} - \bar{X})/\sqrt{S}$ and $(X_{(n)} - \bar{X})/\sqrt{S}$ as a bivariate analogue of (2.9). The density is

$$f_n(x, y) = kD^{n+v-5} F_{n-2}\left\{\frac{(n-1)x+y}{(n-2)D}, \frac{x+(n-1)y}{(n-2)D}\right\} \tag{3.11}$$

where $k = n(n-1)\left(\dfrac{n}{n-2}\right)^{1/2}\left(\dfrac{n+v-3}{2v}\right)$

and $D^2 = 1 - \{(n-1)(x^2+y^2) + 2xy\}/(n-2)$

Tables of fractiles of B and B^* are to be found in Appendix 1.

The statistic B, as was mentioned in Chapter 2, is optimal in the sense of maximizing the probability of correct identification of an outlier when one is present. The model for this is that the outlier is also normally distributed with the same variance but a larger mean than the good data. The test based on B^* may be regarded as an unbiased variant, suitable when it is not known *a priori* whether the contaminating mean is larger or smaller.

Another situation leading naturally to B^* arises if it is believed that the contaminating distribution has a larger variance than the basic distribution, but the same mean.

Using the location-shift model, Ferguson (1961a) showed that the statistics maximizing the power function locally in the neighbourhood of H_0 are the skewness $\Sigma\{(X_i - \bar{X})/s\}^3$ if the sign of the displacement is known, and the kurtosis $\Sigma\{(X_i - \bar{X})/s\}^4$ if it is unknown. As the notation implies, these statistics are for the case $U = v = 0$, and it is not clear how to adapt them to the possibly embarrassing gift of extra information on σ. In fact, as subsequent power studies have shown, despite their local optimality, they are not to be preferred to B and B^* if a single outlier is suspected. They do, however, have a great advantage over the latter statistics of being powerful against several outliers, as well as only one outlier.

3.3 Non-optimal statistics

The discussion above has concentrated on the optimal statistics B and B^*. Apart from these, a large variety of test statistics has been proposed for normal data. The sheer number and variety of these make it an impossible task to document them all, even if one felt the exercise to be worth while. Hence we shall just sketch out a few of the other more significant test statistics proposed.

First, there is the statistic, analogous to B, $(X_{(n)} - \bar{X})/\sqrt{U}$. This test was proposed as an extension of the statistic $(X_{(n)} - \bar{X})/\sigma$ to the case where σ was unknown, but the external information U on σ was available. Its null distribution has been studied by several writers. Nair (1948) deduced its distribution from that of $(X_{(n)} - \bar{X})/\sigma$ using general results on studentized statistics due to Hartley (1938). David (1956) and Halperin *et al.* (1955) approached the problem by the use of the first Bonferroni inequality.

This statistic is inferior to B in that it ignores the information about σ contained in the sample itself. The extent of the inferiority clearly depends very strongly on the magnitude of n and ν, and may be ignored safely if ν is large. A practical advantage of its use is that it is much less sensitive to masking than is B; since, however, masking is a much less serious problem when ν is large, it seems that there is little good reason for using it in preference to B.

A second class of non-optimal statistics is the class based on gaps $X_{(i+1)} - X_{(i)}$. These tests are attractive if we have reason to believe that the contaminants all follow the same distribution. As a single-outlier test, this procedure tests $(X_{(n)} - X_{(n-1)})/\sigma$, with σ being replaced by an estimate if unknown. The null distribution was given by Irwin (1925), who also proposed the statistic $(X_{(n-k+1)} - X_{(n-k)})$ to test for exactly k outliers. Irwin did not make clear what one should do if k were unknown, but one assumes that he would look for the largest gap as an indicator of how many outliers were present.

The latter use of the gaps has been proposed again by Tietjen and Moore (1972) and Tiku (1975). These writers propose that one use the gaps as an indicator of how many outliers to test for, and then apply some other more optimal test specific to that number. As yet, little is known about the effect of doing this on the probability of type I error in the subsequent test, though Tietjen and Moore suggest that this probability is approximately doubled.

A more complicated use of the gaps is in Tukey's (1949) 'gap-straggler' procedure for following up an analysis of variance. Nothing seems to be known about the null probabilities associated with the procedure.

An entirely different class of criteria is the Dixon (1950) r-criteria. These criteria are ratios of the form

$$(X_{(n-i+1)} - X_{(j)})/(X_{(n-k+1)} - X_{(m)})$$

Various different choices of the parameters i, j, k and m may be made, depending on what type of contamination is suspected. The r-criteria seem to have lost favour. Two possible reasons are that they do not extend easily to such situations as the linear model, and their non-optimality. Their early attraction was their relative insensitivity to masking, but this property is shared by the skewness and kurtosis statistics, and is in any event a less severe problem now because of the advent of adaptations of B which are better suited to multiple outliers.

Apart from these statistics which assume normality but are not fully efficient, there are also non-parametric approaches to outliers. These are dealt with in a separate chapter.

3.4 Performance of the outlier tests

The power of an outlier test requires careful definition. Suppose the sample contains a single contaminant $- X_1$; say

$$X_2, \ldots, X_n \sim \mathrm{N}(\xi, \sigma^2)$$

while X_1 follows some contaminating distribution. A general model for this is

$$X_1 \sim \mathrm{N}(\xi + \delta\sigma, \lambda^2\sigma^2) \qquad \delta \geqslant 0, \lambda \geqslant 1$$

from which the case of location contamination emerges as the special case $\lambda = 1$, and scale contamination as the special case $\delta = 0$.

In Chapter 2, a variety of criteria for measuring the performance of outlier tests was mentioned. We will restrict attention to just three of these: β_1, the probability of correct decision; β_2 the power function; and $\beta_4 = \Pr[b_1$ in the critical region].

While β_4 is itself not a useful measure of the performance of the test, it is a good approximation to β_1, and has the advantage of being quite easily evaluated analytically. To verify the first claim, note that

$$\begin{aligned} \beta_1 &= \Pr[b_1 > B_0 \quad \text{and} \quad b_j < b_1] \\ &= \Pr[b_1 > B_0] - \Pr[b_1 > B_0 \quad \text{and some} \quad b_j > b_1] \end{aligned}$$

Now if B_0 exceeds $0.5^{1/2}$, then the Pearson–Chandra Sekar limit shows that the latter probability is zero. Even if B_0 does not exceed $0.5^{1/2}$, it is apparent that the probability must be small. Specifically, $\Pr[b_j > b_1 | b_1 > B_0]$ must be smaller than in the null case since b_j is stochastically smaller than in the null case because of the inflation in S, while b_1 is stochastically larger.

David and Paulson (1965) give the inequality

$$\beta_4 \leqslant \beta_1 \leqslant \beta_4 + (n-1)\gamma$$

where $\gamma = \Pr[b_j > B_0]$.
Hawkins (1974) shows that γ is approximately given by

$$\gamma \approx \Phi\left[-B\frac{\{(f+\tau)/f - (f+2\tau)/2f(f+\tau)\}}{B_0[1 + B_0^2(f+2\tau)/\{2f(f+\tau)\}]^{1/2}}\right] \tag{3.12}$$

where $f = n + v - 2$

Φ denotes the cumulative distribution function of N(0, 1)

$\tau = \delta^2/(n-1)$

it being assumed that $\lambda = 1$; that is, that the contamination is in location only.

If τ is large, the argument behaves like $-B_0\{\tau/(f + B_0^2)\}^{1/2}$ and clearly γ tends to zero very rapidly as τ increases. Simple manipulations of this expression show that γ decreases rapidly as β_4 increases. Thus a simple, but much weaker inequality is

$$\beta_4 \leqslant \beta_2 \leqslant \beta_4 + \alpha \qquad \text{(cf. Equation (2.11))}$$

as is also shown in David and Paulson (1965).

While general inequalities connecting β_1 and β_4 seem not to have been derived, a number of tabulations in Hawkins (1969b) suggest that

$$\beta_4(1-\alpha) < \beta_1 \leqslant \beta_4$$

These inequalities make it clear that while there is a considerable scope for debate about which measure is best in a particular context, the measures differ negligibly from one another provided α is in the usual range of 0.1 and under. It is also a consequence of the inequalities that any test which is optimal under one of the measures will be near optimal under another.

For the model of location contamination only, performance data have appeared in a number of sources – Dixon (1950), David (1956), Ferguson (1961b).

Figure 3.1 provides a good bird's-eye view of the effect on β_4 of varying n, ν and α. The figures in fact show, not only the effect of varying n, ν and α; but also that of using a non-optimal test statistic. The curves shown solid are for the Nair statistic, while the dashed curves are for B. Comparing them shows quite clearly how large is the benefit of increasing ν when n is small. It is also clear that the Nair statistic is most inferior to B when ν and n are both small; both these

(i)

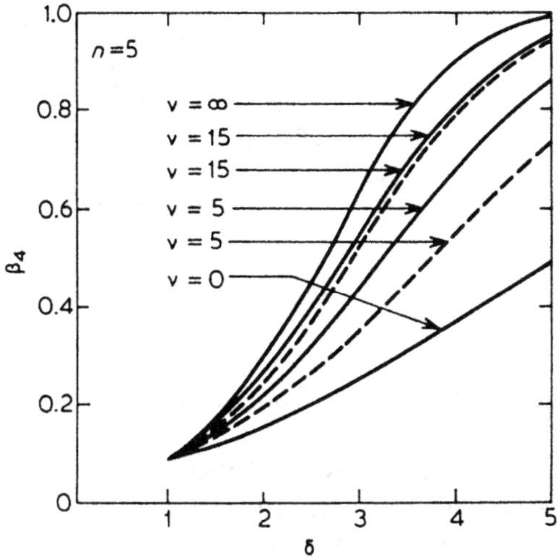

(ii)

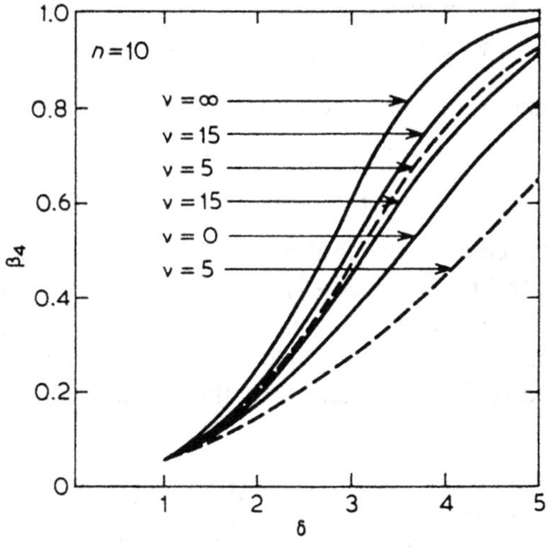

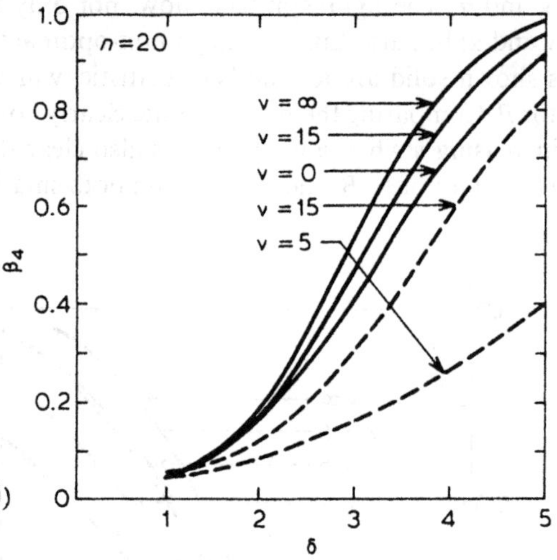

(iii)

Figure 3.1

conclusions are to be expected from the qualitative discussion given earlier.

Alternatively, following Hawkins (1969b), one may note that by the Bonferroni approximation, the critical region $B > B_0$ is exactly or nearly exactly the same as max $t_i > T_0$ where T_0 is the α/n fractile of a Student's t-distribution with $n+v-2$ degrees of freedom. If the model of the contaminant is

$$X_1 \sim \mathrm{N}(\xi + \delta\sigma, \lambda^2\sigma^2) \qquad \delta \geqslant 0, \lambda \geqslant 1$$

then, as some elementary operations show,

$$t' = \left\{\frac{(n-1)(n+v-2)}{(n-1)\lambda^2+1}\right\}^{1/2}(X_1 - \bar{X}_1)/\surd S_1$$

follows a noncentral t-distribution with $n+v-2$ degrees of freedom and noncentrality $\Delta = \delta[n-1)/\{(n-1)\lambda^2+1\}]^{1/2}$ and the probability β_4 reduces to $\Pr[t' > T_0\{n/([n-1]\lambda^2+1)\}^{1/2}]$.

Using the Johnson and Welch (1939) approximation to the cumulative distribution function of the noncentral t, β_4 is approximately

$$\beta_4(\delta) \approx \Phi\left[\frac{\delta(n-1)^{1/2} - T_0 n^{1/2}}{\{(n-1)\lambda^2 + 1 + \frac{1}{2}nT_0^2/(n+v-2)\}^{1/2}}\right] \qquad (3.13)$$

This approximation is rather rough if $n+v$ is not large, but it is more than adequate for our purposes. It shows for example that 50 per cent power is attained when δ is about equal to T_0, the fractile used in the test. For fixed δ, any increase in λ causes β_4 to move towards 0.5. Since T_0, the α/n fractile of a t-distribution with $n+v-2$ degrees of freedom, is not especially sensitive to the value of v provided $n+v-2$ is large, we see that an increase in v does not lead to a dramatic improvement in power unless n, v and λ are all small.

The power of the test based on B^* may be approximated in the same way. The critical region $B^* > B_0^*$ is exactly or nearly exactly equivalent to $\max|t_i| > T_0^*$ where T_0^* is the $\alpha/2n$ fractile of a Student's t-distribution with $n+v-2$ degrees of freedom. It is easily verified that the approximate expression for β_4 becomes

$$\beta_4 = \beta_4(\delta) + \beta_4(-\delta), \qquad \text{with} \qquad \beta_4(\delta) \text{ as before}$$

and an armchair analysis like that for B may be carried out quite easily.

The other measures of performance β_1 and β_2 do not lend themselves to quite such easy analysis. Exact expressions for β_1, together with some tables, are to be found in Hawkins (1969b). Figures for β_2, and for criteria other than B, are generally much harder to find, and have for the most part been produced by means of simulation studies.

Tables 3.1 and 3.2 (taken from Hawkins, 1969b) illustrate the relative magnitudes of β_1, β_4 and β_3, and may also be used to confirm the accuracy of the approximate expression given. In the first case, β_1 and β_4 are identical, while in the second, their values agree to within

TABLE 3.1 Powers of B. $n=5$, $v=0$, $\alpha=0.05$

λ	δ	0	2	4	5	8
1	β_1, β_4	0.010	0.114	0.404	0.730	0.921
	β_3	0.050	0.145	0.408	0.731	0.921
3	β_1, β_4	0.097	0.232	0.431	0.647	0.821
	β_3	0.262	0.372	0.518	0.683	0.831
5	β_1, β_4	0.193	0.311	0.452	0.598	0.732
	β_3	0.461	0.540	0.626	0.714	0.799
7	β_1, β_4	0.261	0.357	0.464	0.573	0.676
	β_3	0.590	0.644	0.699	0.753	0.806
9	β_1, β_4	0.306	0.386	0.471	0.557	0.641
	β_3	0.674	0.711	0.749	0.796	0.822

TABLE 3.2 Powers of B. $n = 15$, $v = 15$, $\alpha = 0.05$

λ	δ	0	2	4	6	8
1	β_4	0.003	0.183	0.815	0.996	1.00
	β_1	0.003	0.178	0.806	0.995	1.00
	β_3	0.050	0.294	0.825	0.996	1.00
3	β_1	0.161	0.370	0.628	0.838	0.950
	β_1	0.160	0.368	0.625	0.836	0.949
	β_3	0.554	0.682	0.808	0.908	0.968
5	β_4	0.275	0.421	0.578	0.724	0.840
	β_1	0.273	0.419	0.577	0.723	0.839
	β_3	0.744	0.800	0.853	0.900	0.938
7	β_4	0.334	0.443	0.556	0.665	0.762
	β_1	0.333	0.442	0.555	0.664	0.761
	β_3	0.824	0.855	0.884	0.910	0.934

0.01 throughout the table. It is of interest that β_3 tends to 1 as $\lambda \to \infty$, while both β_1 and $\beta_4 \to 0.5$. The significance of this is that if λ is large, then B is not the appropriate statistic, and one should use B^*. For $\lambda = 1$ when B is the correct statistic, the value of β_3 is not terribly different from those of β_1 and β_4, and it is of interest that the maximum discrepancy seems to occur in the region near H_0, where there is a real danger that it is not the actual contaminant that is being tested by B.

3.5 The use of normal approximations

It is quite common practice to invoke the asymptotic normality of many families of distributions (for example the gamma, Poisson, binomial) to justify the use of the normal distribution as an approximation in non-asymptotic cases. These approximations unfortunately tend to be poorest in the extreme tails of the distribution – the exact area of most importance for outlier testing. Thus a normal approximation which is excellent for most purposes can turn out to be very poor for purposes of outlier testing. Consider for example the chi-squared distribution, and let $X_i \sim \chi^2_v, i = 1, \ldots, n$. As is well known, if v is large,

$$X_i \sim \mathrm{N}(v, 2v)$$

and to a better degree of approximation

$$\surd(2X_i) \sim \mathrm{N}(\surd\{2v - 1\}, 1)$$

For $\nu = 30$, these two normal approximations yield approximations to the upper 0.01 fractile whose right tail areas are 0.020 and 0.0122 respectively. For $\nu = 50$, these figures improve to 0.018 and 0.0118 respectively. While the first approximation seems too rough, the second is excellent.

However, if the approximations are used to estimate the upper 0.01 fractile for the largest of $n = \nu$ such chi-squared variates, we find that for $n = \nu = 30$, the first approximation yields an actual tail area of 0.40 amd the second, 0.02. For $n = \nu = 50$, these figures become 0.07 and 0.02; clearly not very satisfactory.

In fact as $n \to \infty$, extreme-value theory tells one that

$$X_{(n)} = 2\log n + o(1) \qquad (3.14)$$

The crude normal approximation would have it that

$$X_{(n)} = \sqrt{(4\nu \log n)} + \nu + o(1) \qquad (3.15)$$

while the second, more accurate approximation implies that

$$X_{(n)} = \log n + \tfrac{1}{2}\sqrt{\{(4\nu - 2)\log n\}} + \nu - \tfrac{1}{2} + o(1) \qquad (3.16)$$

and both are quite badly in error.

A similar analysis could be made for many other approximately normal distributions, but such a detailed compendium would add little insight to the discussion. It is thus clear that unless the normal distribution approximates the actual distribution in the extreme tails very well, the conclusions drawn from use of the normal approximation can be wrong to an almost arbitrarily large extent. Thus normal approximations should be used with extreme caution, if at all.

CHAPTER 4

The gamma distribution

4.1 Introduction

The general form of the gamma distribution is

$$f(x|\alpha, \beta) = \beta^{\alpha} x^{\alpha-1} e^{-\beta x}/\Gamma(\alpha) \qquad x \geqslant 0$$

The parameter α is a shape parameter and β a scale parameter.

The gamma distribution arises in two quite common ways. One is the chi-squared distribution. Here, 2α is (commonly) an integer – the number of degrees of freedom, while $\beta = 0.5$. It is this model that usually gives rise to outlier problems for the gamma distribution, and much of the discussion will centre on the homoscedasticity problem arising from analysis of variance.

The other genesis of the gamma distribution arises in the Poisson process with parameter β. The distribution of the time to the αth event is the gamma, and so for this problem, α must be an integer. While the use of an outlier test would be an interesting and novel method of testing a Poisson process for constancy of β, no-one seems to have actually done it.

A slightly different situation arises when we have a number of gamma variates to be compared with a single reference variate. This problem will be deferred for the moment.

A common framework giving rise to gamma outliers arises in analysis of variance. We illustrate this with a one-way analysis of variance, though the idea is perfectly general.

Let $\qquad Y_{ij} \sim N[\xi_i, \sigma_i^2] \qquad j = 1 \text{ to } m_i,\ i = 1 \text{ to } n$

Let $\qquad Y_{i.} = \sum_j Y_{ij}/m_i$

$$X_i = \sum_j (Y_{ij} - Y_{i.})^2$$

Then $X_i \sim \sigma_i^2\,\chi^2_{\nu_i}$, $i = 1$ to n, where $\nu_i = m_i - 1$ ($= 2\alpha$ in our earlier notation) and we assume all $m > 1$ (if this is not true for a particular i, then the X_i for that cell is zero and so is uninformative. We remove it, decreasing n by 1).

The model for analysis of variance requires that all σ_i be equal. One possible alternative to this is the 'slippage hypothesis'. In its simplest form, this is that one σ_i is larger (or smaller) than the remainder, which are all equal.

The easiest case occurs when all ν_i are equal. The β_1 optimal test for slippage to the right (one σ_i larger than the others) is deducible from Karlin and Truax's (1960) general theory, and is $X_{(n)}/W$, where $W = \Sigma_i X_i$. Rejection is for large values of this ratio. The test for slippage to the left rejects H_0 on small values of $X_{(1)}/W$.

The statistic $X_{(n)}/W$ is due to Cochran (1941) and $X_{(1)}/W$ to Doornbos (1956). The exact distribution of $X_{(n)}/W$ was found by Hawkins (1972) and is another special case of the general recursion (2.9). The method can be adapted very easily to find the distribution of $X_{(1)}/W$ also.

In fact it will be convenient to derive the distribution of a slightly more general statistic: assume that we may also have to hand some additional information about σ in the form of V, an independent $\sigma^2\chi^2$ variable with m degrees of freedom. As usual, if no such variable is to hand, then we set $V = m = 0$ in the general results. With this convention in mind, we redefine W as $W = V + \Sigma_1^n X_i$.

Without loss of generality, we may assume that $\sigma = 1$. Then letting $F_n(.)$ denote the distribution function of $X_{(n)}/W$, the following recursive expression may be derived. The density of $S = X_i/W$ for some arbitrary i is

$$q(s) = \frac{\Gamma[\frac{1}{2}(\nu n + m)]\, s^{\nu/2 - 1}(1 - s)^{1/2\{\nu(n-1)+m\}-1}}{\Gamma(\frac{1}{2}\nu)\Gamma[\frac{1}{2}\{\nu(n-1) + m\}]} \tag{4.1}$$

then

$$F_n(s) = n \int_0^s g(t) F_{n-1}\{t/(1-t)\}\,\mathrm{d}t \tag{4.2}$$

$$F_0(s) = \begin{cases} 0 & s < 0 \\ 1 & s \geqslant 0 \end{cases}$$

and so the general distribution may be found recursively. Similarly, the exact cumulative distribution function of $X_{(1)}/W$ – say $F_n^*(.)$ – is given by

$$F_n^*(s) = n \int_0^s g(t)[1 - F_{n-1}^*\{t/(1-t)\}]\,dt \tag{4.3}$$

using the starting values $F_0^*(s) = 1$, $s \geqslant 0$.

In the case $\nu = 2$, $m = 0$, the X_i are simply scaled exponentially distributed variables, and then $Z_i = X_i/W$ follows a uniform distribution, and the set $X_{(i)}/W$ is distributed like the order statistics of a uniformly distributed sample. Using this fact, Wani and Kabe (1973) show how the distribution of a general class of statistics including both those discussed here may be derived. The expressions resulting have the desirable property of being in closed form. Unfortunately, if the common value of ν is not 2, then the set of Z follows a Dirichlet distribution, and their method becomes more complicated than the recursive method above.

The Bonferroni approximations to the fractiles of $X_{(n)}/W$ and $X_{(1)}/W$ are quite simple. Note that by the independence of the X_i, under the null hypothesis $X_i \sim \chi_\nu^2$ and $W - X_i \sim \chi_{\nu(n-1)+m}^2$ are independent. Thus

$$\frac{X_i}{W - X_i} \frac{\nu(n-1) + m}{\nu} \sim F_{\nu, \nu(n-1)+m} \tag{4.4}$$

Letting k_α be the upper α/n fractile of the $F_{\nu,\nu(n-1)+m}$ distribution, we then have

$$\frac{X_i}{W - X_i} \frac{\nu(n-1) + m}{\nu} < k_\alpha \Leftrightarrow \frac{X_i}{W} < K_\alpha = \frac{\nu k_\alpha}{\nu(n - 1 + k_\alpha) + m}$$

Thus

$$\Pr[X_{(n)}/W > K_\alpha] = \Pr\left[\bigcup_1^n X_i/W > K_\alpha\right]$$

$$\leqslant \sum_1^n \Pr[X_i/W > K_\alpha] = \alpha \tag{4.5}$$

The Bonferroni approximation to the α fractile is exact if K_α is such that $(X_i/W > K_\alpha) \cap (X_j/W > K_\alpha)$ is an impossible event for all pairs i and j. Obviously, since $W \geqslant X_i + X_j$, this occurs when $K_\alpha > 0.5$. This condition is frequently met for moderately small values of both ν and n.

In an exactly analogous way, an approximation to the fractiles of $X_{(1)}/W$ may be derived. We may note in passing that this approximation is never exact for $n > 2$, although as Doornbos (1959) shows, the size of the test based on it lies between $\alpha - \alpha^2/2$ and α; a very narrow range if α is small.

The distributional result just outlined are in terms of the familiar F-distribution. Alternatively, we may note the equivalent fact that X_i/W follows a type-1 beta distribution, Equation (4.1), from which the identical results may be derived.

4.2 The problem of unequal v_i

If the v_i are not all equal, then the X_i are no longer exchangeable and the general theories of optimal tests can no longer be invoked. Thus for this situation, no optimal test is known.

There are two intuitively appealing approaches to the problem. In the first, for each i, evaluate a_i: the significance of the ratio X_i/W. This ratio is equivalent to

$$\frac{X_i}{W - X_i} \frac{v_i}{N - v_i} \qquad N = \sum_1^n v_i + m$$

which follows an F-distribution with v_i and $N - v_i$ degrees of freedom. If the smallest of the a_i is less than α/n, then reject the corresponding X_i. The choice of α/n is, of course, made on the basis of Bonferroni's inequality. The alternative approach is to set up a tolerance region for each X_i using the remaining X_j. We reject the null hypothesis if any X_i does not lie within its tolerance region. In fact, choosing the content of the tolerance regions to yield an overall probability of type I error of α yields a procedure identical to that just described.

4.3 Performance of the tests

Suppose now that $X_1 \sim \lambda^2\sigma^2\chi_v^2$ $(\lambda > 1)$ while $X_2, \ldots, X_n \sim \sigma^2\chi_v^2$. The probability of a correct identification of X_1 as an outlier is given in Hawkins (1972): let $g^*(.)$ denote the beta type-2 density of $(W - X_1)/X_1$ and suppose that K is the fractile of $X_{(n)}/W$ used in the test. Then

$$\beta_1 = \int_0^{K^{-1}-1} g^*(Y)F_{n-1}(Y^{-1})\mathrm{d}Y \tag{4.6}$$

As yet, no general exact tables of β_1 have been published. An approximation to the probability of correct decision is

$$\beta_4 = \Pr[X_1/W > K]$$

Clearly, this overstates the real probability of correct decision. Specifically,

$$\beta_1 = \beta_4 . \Pr[X_j < X_1 \quad \text{all} \quad j \neq 1 | X_1/W > K]$$

However: (i) if $K > 0.5$, this conditional probability is 1, and then $\beta_1 = \beta_4$; (ii) even when $K < 0.5$, the conditional probability must exceed $1 - \alpha$, and so here β_4 is a good approximation to β_1. To see that the probability cannot be less than $1 - \alpha$, we note two facts: first that the conditional probability must be decreasing function of λ for $\lambda > 1$, and second that for $\lambda = 1$,

$$\Pr[X_j > X_1 | X_1/W > K] \leqslant \frac{\Pr[(X_j/W > K) \cap (X_1/W > K)]}{\Pr[X_i/W > K]}$$

$$< \alpha^2/\alpha = \alpha \qquad \text{(Doornbos, 1956)}$$

Now the event $X_1/W > K$ is equivalent to

$$\frac{X_1}{W - X_1} \, \frac{\nu(n-1) + m}{\lambda^2 \nu} > \frac{K\{\nu(n-1) + m\}}{\lambda^2 \nu (1 - K)}$$

and the random variable on the left follows an F-distribution with ν and $\nu(n-1) + m$ degrees of freedom.

Thus the (central) F-distribution may be used to give an excellent approximation to the probability of correct decision.

4.4 Comparison with the Bartlett test

Suppose again that $X_i \sim \sigma^2 \chi^2_{\nu_i}$, $i = 1$ to n. The common test for H_0 : $\sigma_1 = \sigma$, all i, is Bartlett's test, a modification of the generalized likelihood ratio test. The alternative hypothesis is simply that H_0 is false – not all σ_i are equal. The statistic is

$$B = c\left[\sum_i^n \nu_i \log(X_i/W) + m \log(V/W)\right] \qquad (4.7)$$

It is well known that B is not a robust statistic – significant values may as easily indicate non-normality as differences amongst the σ_i. Note in particular that B is not robust against situations in which there is a perceptible increase in the probability of some X_i being near zero. This situation is very common in real data as a consequence of data only being recorded to a limited number of significant digits. For example, a set of data may behave very much like normal data except that only integer values are observed. If this occurs, the relatively minor perturbations of the density of X_i near zero may make B useless. On the other hand $X_{(n)}/W$ is robust against this kind of deviation from model. Of course, neither statistic is robust against a distribution that actually has a right tail that is too heavy.

This suggests that for some data, the Cochran statistic is more robust than Bartlett's. Thus, if the alternative hypothesis is (or approximates) a slippage model, the use of Cochran's test may lead to an improvement in both power and robustness. Further, an extension of Cochran's statistic to deal with multiple outliers may lead to powerful robust procedures for more general alternatives.

No full study of the relative power and robustness of the Cochran and Bartlett statistics seems to have been carried out. A small simulation was carried out to illustrate the points made earlier. Two values of v, 5 and 10, were used with $n = 10$, $m = 0$ and 5000 random samples drawn. From these, the fractiles and powers were estimated. (We note that on the face of it, it is not necessary to estimate these fractiles, since we have to hand the excellent Bonferroni approximation for the null fractiles of the Cochran statistic, and the asymptotic χ^2 distribution of the Bartlett statistic. The use of fractiles estimated from the data was made to avoid the possibility of doubts about the appropriateness of the asymptotic theory clouding the interpretation of the performances.)

The estimated powers are set out in Table 4.1. The superior power of the Cochran statistic is evident. In an attempt to assess the robustness of the two procedures against a distribution which is heavier near zero than the gamma, data were generated as

$$X = Y[1 - 0.5(v/\{v + Y\})^2]$$

TABLE 4.1

	$n = 10$			
	$v = 5$		$v = 10$	
λ	Cochran	Bartlett	Cochran	Bartlett
2	0.48	0.32	0.74	0.59
3	0.83	0.72	0.98	0.95
4	0.94	0.88		
5	0.98	0.95		

TABLE 4.2

	n = 10			
	$\nu = 5$		$\nu = 10$	
Nominal	Cochran	Bartlett	Cochran	Bartlett
0.1	0.16	0.26	0.18	0.25
0.05	0.10	0.15	0.11	0.15
0.01	0.03	0.05	0.03	0.05

where Y is a χ_ν^2 variate. For large values, X is effectively a gamma variate, but small values are displaced towards the origin.

Table 4.2 shows the estimated size of the tests using the nominal 0.1, 0.05 and 0.01 upper fractiles found in the first part of the simulation.

The greater robustness of the Cochran statistic is apparent – the error in the actual size of the test is only about half that of Bartlett's statistic.

In addition to the obvious applications of this theory in analysis of variance for testing for homoscedasticity, there are some more unusual ones. One relates to the analysis of a 2^p factorial experiment (Birnbaum, 1959). Here, if most of the effects and interactions are believed to be zero, a slippage analysis of the $2^p - 1$ squares of contrasts may lead to a more effective analysis than the usual method of supposing *a priori* that all the contrasts in a certain set are not significant, and using them to estimate the residual variance. This idea of course just formalizes the use of a half-normal plot.

Another slightly non-standard application is to stepwise regression using orthogonal predictors. X_i is the variance explained by predictor i.

Little seems to be known about the statistic for slippage to the left, $X_{(1)}/W$. As noted earlier, conservative fractiles may be set up by means of Bonferroni's inequality: if k denotes the lower α/n fractile of an F-distribution with ν and $m + \nu(n-1)$ degrees of freedom (this is the inverse of the $1 - \alpha/n$ fractile of an F-distribution with $m + \nu(n - 1)$ and ν degrees of freedom), and $K = k\nu/\{\nu(n-1+k)+m\}$, then

$$\Pr[X_{(1)}/W < K] \leqslant \alpha$$

and an approximation to the power may be found in the same way as in the preceding section using the (central) F-distribution.

From the very little attention given to the hypothesis of slippage to the left, one might conclude that it is much less common than slippage to the right. A more probable explanation, however, is that a single variance which has slipped to the left will lead to a relatively minor downward bias in a pooled variance estimate – a very large downward bias is precluded by the fact that a gamma variate is non-negative. As against this, a variance that has slipped to the right may exert an arbitrarily large effect.

It is not entirely trivial to set up a test for the hypothesis that one variance has slipped, but the direction of the slip is unspecified. An obviously 'sensible' procedure is to accept H_0 if neither $X_{(1)}/W$ nor $X_{(n)}/W$ is significant. Suitable conservative fractiles may be found by using the $\alpha/2$ fractile of each statistic. If the null hypothesis is rejected then one rejects the more significant of $X_{(n)}$ and $X_{(1)}$.

The optimality of this procedure seems not to have been proved, but it is unbiased, and inherits the optimality properties of the Cochran and Doornbos statistics.

Another statistic is $X_{(n)}/X_{(1)}$, which is clearly a good statistic to use if one believes that there is slippage in more than one observation in opposite directions. It is presumably optimal against the alternative of two outliers – one to the right and one to left.

4.5 The maximum *F*-ratio

Another related problem to the gamma slippage problem is the following one: given $X_1, X_2, \ldots, X_n$ believed to be $\sigma^2\chi_v^2$ variates, and V, a $\sigma^2\chi_m^2$ variate, test whether any of the ratios X_i/V is significantly large. This problem arises quite naturally in a multiway analysis of variance, where V is the 'error' sum of squares and $\{X_i\}$ the sums of squares corresponding to the n hypotheses being tested.

We have already seen a particular example of this type of situation as it arises in stepwise regression. The statistic used there was $X_{(n)}/W$. The more common statistic used in analysis of variance, on the other hand, is the different statistic

$$F_{(n)} = mX_{(n)}/(Vv)$$

This is the largest of the n F-ratios $mX_i/(Vv)$: hence the name 'maximum F-ratio'.

The distribution of $F_{(n)}$ may be evaluated from that of $X_{(n)}$ in either of two ways: by setting up the bivariate distribution of $X_{(n)}$ and V and applying standard transformation theory: or by using Hartley's (1938) asymptotic expansion of the distribution of studentized statistics. If v is an even integer, then the distribution of $X_{(n)}$ is relatively straightforward and no real difficulties other than algebraic tedium arise. Fisher (1929) derived the distribution for $v = 2$, and Finney (1941) extended this to arbitrary even integers v and m.

Tables of some fractiles of $F_{(n)}$ are to be found in Pearson and Hartley, (1954). A question arises quite naturally of which of $F_{(n)}$ and $X_{(n)}/W$ is to be preferred. General considerations show that the latter statistic must be more powerful if it is suspected that most of the X_i do have the same scale parameter as V, for then the estimator W of this scale parameter is preferable to V.

Of course, this benefit is lost if enough of the X_i are outlying, and one then has the problem of carrying out a multiple outlier test on the data. In this case, the robustness of $F_{(n)}$ to multiple outliers amongst the X_j makes it more attractive. This situation is closely analogous to that found in the case of the normal distribution. On the one hand, one wishes to use the statistic which is optimal for a single outlier. On the other, this statistic is susceptible to masking if there are in fact several outliers, and so the choice is not entirely clear-cut.

Clearly, much still needs to be done to determine when each of the statistics is to be preferred. That the controversy may take long to resolve is suggested by the related controversy over whether, having found an interaction term in an analysis of variance to be non-significant, one should pool it with the error variance.

CHAPTER 5

Multiple outliers

5.1 Introduction

Up to now, it has been assumed tacitly that at most one outlier was present in the sample. Under the contamination model, if the expected number of outliers in the sample, np, is small, this model may be perfectly reasonable as a first approximation. However, it must be realized that most practical problems may have several outliers, and one should, as a matter of course, use test procedures that are effective with multiple outliers.

An obvious, and early, approach is the following: test $X_1, \ldots, X_n$ for a single outlier. If none is found, conclude that the entire sample is 'clean'. If an outlier is found, remove it from the sample and repeat the testing procedure on the remaining observations. That this procedure has a grave defect was shown by Pearson and Chandra Sekar in 1936. To illustrate the problem, suppose that $X_1, \ldots, X_n$ is a sample believed to be N(0, 1). To the top two order statistics, $X_{(n)}$ and $X_{(n-1)}$, add a deviation δ. Now as $\delta \to \infty$ we can easily verify that

$$\bar{X} = 2\delta/n + o(1)$$

and

$$\begin{aligned} s^2 &= \sum_1^n (X_i - \bar{X})^2 \\ &= (n-2) \cdot 4\delta^2/n + 2(\delta - 2\delta/n)^2 + o(1) \\ &= 2(n-2)\delta^2/n + o(1) \end{aligned}$$

Thus

$$(X_{(n)} - \bar{X})/s \to \{(n-2)/2n\}^{1/2} = C, \text{ say} \tag{5.1}$$

Now suppose the fractile used for testing the statistic exceeds C. Then as $\delta \to \infty$, the probability of rejecting $X_{(n)}$ goes to zero: paradoxically, the more outlying the pair $X_{(n)}$, $X_{(n-1)}$, the less the probability of identifying $X_{(n)}$ as an outlier.

This effect occurs quite generally – as a class, the statistics that are effective in identifying a single outlier tend to lose power badly if more than one outlier is present. The term 'masking effect' has been coined for the tendency. The loss of power occurs for all sample sizes n, though it is only for small n that it is severe enough to reduce the power of the test asymptotically to zero.

It must be supposed that the prevalence and potential seriousness of the masking effect is the main cause, both of the large degree of attention given in the literature to the detection of multiple outliers, and the fact that none of the solutions proposed is entirely satisfactory. A general outline of the current state of the art would be as follows. As practically all material written relates to the normal distribution with unknown mean and variance, this situation is assumed implicitly.

5.1.1 *Number and location of outliers known*

Considerably the easiest case arises when one knows *a priori* both how many contaminants there are, and how many lie on the left, and on the right of the sample. Even here, however, the problem is not entirely clear-cut. First, if we assume that the full distribution of the sample is $X_{j(i)} \sim N[\xi + a_i\delta, \sigma^2]$, where a_i are known, $j(i)$ is an unknown permutation and $a_1 \geqslant a_2 \geqslant \ldots \geqslant a_n$, then by the theory mentioned in Chapter 2, the optimal test for $\delta = 0$ uses the general Murphy statistic as its test statistic:

$$\sum_{i=1}^{n} a_i(X_{(i)} - \bar{X})/\sqrt{S} \tag{5.2}$$

In particular, if $\Sigma a_i = 0$, the optimal statistic is $\Sigma a_i X_{(i)}/\sqrt{S}$. This result generalizes the result mentioned in Chapter 2 that if two contaminants are suspected, and have means believed to be equally spaced above and below ξ, then the studentized range is the best statistic.

In the more usual situation, it is not known what the a_i are. Here there is no optimality theory to guide us, but the following procedure is reasonable:

If one anticipates m outliers on the left and r on the right, let $X_{(m+1)}$, $X_{(m+2)}, \ldots, X_{(n-r)}$ be the central order statistics. Let $\bar{X}_{m,r}$ be their mean, and

$$s^2_{m,r} = \sum_{i=m+1}^{n-r} (X_{(i)} - \bar{X}_{m,r})^2, \qquad S_{m,r} = s^2_{m,r} + U \tag{5.3}$$

Reject H_0 on low values of $S_{m,r}/S$.

This statistic is a slight extension of Tietjen and Moore's L_k statistic: specifically, L_k is the statistic $s^2_{k,0}/s^2$ (note the assumption that $U = v = 0$).

Alternatively, following Tiku (1975), one may use $X_{(m+1)}, \ldots, X_{(n-r)}$ as a censored sample from an assumed common normal distribution, and construct from them $\hat{\sigma}_{m,r}$ – an optimal estimator of σ from the censored sample. The ratio $S/n\hat{\sigma}^2_{m,r}$ then provides a test statistic. We note that the statistic $S_{m,r}/S$ resembles this rather closely, but uses $S_{m,r}$ as the estimator of σ^2. It may be shown that asymptotically, if m and r remain fixed while $n \to \infty$, $\hat{\sigma}^2_{m,r} \to cS_{m,r}/n$, where c is a constant independent of m and r.

A difficulty with all of these statistics is that general exact distributions are known for none of them except in very special cases. The Murphy and Tietjen–Moore statistics reduce to single outlier ones when $m = 0$, $r = 1$, or $m = 1$, $r = 0$, and they are equivalent. The exact distributions then follow from the single outlier case.

If $m = 0$, $r = 2$, or $m = 2$, $r = 0$, then the distribution of $s^2_{0,2}/s^2$ for $v = 0$ is given by Grubbs (1950) in the form of a multiple integral. For general v, Hawkins (1978a) derives the joint distribution of

$$B = (X_{(n)} - \bar{X})/\sqrt{S}$$

and

$$R = (X_{(n-1)} - \bar{X}_{0,1})/\sqrt{S}$$

From this distribution, that of $S_{0,2}/S$ follows because of the identity

$$S_{0,2}/S = \{1 - nB^2/(n-1)\}\{1 - (n-1)R^2/(n-2)\} \tag{5.4}$$

which is quite easily proved by means of identities from analysis of variance, which may be found in Quesenberry and David (1961).

This joint density also yields the density of the Murphy statistic

$$\begin{aligned} M &= (X_{(n)} + X_{(n-1)} - 2\bar{X})/\sqrt{S} \\ &= (n-2)B/(n-1) + R\{1 - nB^2/(n-1)\}^{1/2} \end{aligned} \tag{5.5}$$

suitable for the case $a_1 = a_2 = 1$, $a_3 = \ldots = a_n = 0$, that is when one suspects two outliers on the right having the same mean.

In Appendix 5, fractiles of M, R and $S_{0,2}/S$ (denoted there by G) are to be found. If $m = r = 1$, the Murphy statistic for $a_1 = 1$, $a_n = -1$, $a_2 = \ldots = a_{n-1} = 0$ is the studentized range, whose distribution is well known (see for example Pearson and Hartley, 1954).

Since $S_{11}/S = 1 - \{(n-1)(b_1^2 + b_n^2) + 2b_1b_n\}(n-2)$ where $b_1 = (X_{(1)} - \bar{X})/\sqrt{S}$ and $b_n = (X_{(n)} - \bar{X})/\sqrt{S}$, its distribution could, in principle, be evaluated from the joint density of b_1 and b_n, which is given by Equation (3.11). In practice, however, the integration necessary to obtain fractiles of S_{11}/S turns out to involve some tedious computation, which seems to have discouraged any would-be tabulators of fractiles.

More general exact results are not known. For the Tiku family of statistics with $v = 0$, approximate results are provided by asymptotic theory: if u denotes the α fractile of a beta variable with parameters $n - m - r - 1$ and $m + r$, then the α fractile of Tiku's statistic can be approximated by $u(n-1)/(n-m-r-1) + \{1 + 1/(n - 2r + 1)\}/5n$.

The Tietjen–Moore statistic, L_k, has been simulated, and fractiles estimated for $m = 0$, various values of r and $n \leqslant 50$. These estimated fractiles are listed in Appendix 2. The simulation approach seems for the moment to provide the easiest solution to the problem of fractiles of the more general $S_{m,r}/S$ statistics. As an alternative, one may obtain conservative fractiles (but possibly unduly so) by using the fractiles of the E statistic discussed below.

Another possibility for conservative fractiles lies in the use of Bonferroni's inequality. Here we observe that $S_{m,r}/S$ is the smallest of a number M of possible variance ratios that could be computed, and an arbitrary one of which would follows a beta distribution. Thus a conservative fractile may be obtained by using the α/M fractile of this beta distribution. The Bonferroni multiplier M is easily seen to be

$$M = n!/\{m!r!(n-m-r)!\}$$

Unfortunately, this Bonferroni approximation, extremely accurate when $m + r = 1$, is unduly conservative for $m + r > 1$.

Consider, for example, the case $n = 10$, $v = 5$, say. Deleting two observations on the right leads to the Anova:

Source	SS	df
Total	S	$n + v - 1 = 14$
Two outliers	$S - S_{0,2}$	2
Residual	$S_{0,2}$	$n + v - 3 = 12$

for which $F = (S - S_{0,2})/S_{0,2} \cdot (n + v - 3)/2$.

The actual 0.05 fractile of $S_{0,2}/S$ is 0.431, which corresponds to $F = 7.92$. The tail area of this value on the F-distribution with 2 and 12 degrees of freedom is 0.0065 which, multiplied by M, yields 0.29.

Thus, the true fractile for $\alpha = 0.05$ is that that would be obtained using the Bonferroni inequality with $\alpha = 0.29$ – clearly an unduly high degree of conservatism.

A lower bound is provided by Gentleman and Wilk (1975). In a discussion containing rather more generality than is used here, they note that for $v=0$, the $n-1$ degrees of freedom in S may be decomposed into $(n-1)/k$ χ_k^2 terms, each corresponding to an outlier statistic for k of the X_i $(k=m+r)$. Amongst these $(n-1)/k$ terms, let S_L correspond to the smallest of the residual sums of squared deviations. Clearly $S_L \geqslant S_{m,r}$, and so S_L/S provides a stochastic upper bound for $S_{m,r}/S$.

Gentleman and Wilk assume that it is reasonable to regard S_L/S as the smallest of $(n-1)/k$ independent beta variables, which for α small is essentially equivalent to using the Bonferroni approximation with a multiplier of $M=(n-1)/k$. They express the hope that this lower bound will be a good approximation to the fractiles of $S_{m,r}/S$. However, some comparisons of this approximation with actual known fractiles soon dispel this hope. For example, if $n=20$, $v=0$, $k=2$, the $\alpha=0.05$ fractile of $S_{0,2}/S=0.479$, corresponding to an F-value on 2 and 17 degrees of freedom of 9.24, and a tail area of 0.002. This corresponds to a Bonferroni multiplier of 25, against the 9.5 suggested by Gentleman and Wilk, or the conservative 190 arising from normal use of Bonferroni's inequality.

The source of the conservatism of the Bonferroni test is not at all hard to find. If one particular X deviates markedly from $\bar{X}$, then all pairs of X values containing it as one member will tend to yield large F-ratios. This means that the M possible test statistics contain many highly correlated pairs, and so the Bonferroni approximation is poor.

This fact suggests that then the use of the kth approximand in the Boole expansion should yield a much better approximation than the Bonferroni one. This conjecture remains to be verified by actual computation.

From all this, it can be seen that the situation *vis-à-vis* published fractiles for the general $S_{m,r}/S$ statistics (apart from L) is poor. However, this situation is probably not serious in practice since in actual application, unless one knew *a priori* that all outliers were either on the right of the sample or on the left (in which case one would use the L_k statistics), one would be unable to say how many of the outliers should be on the right, and how many of the left of the sample, and so one would use the E_k statistic defined in the next section.

5.1.2 *Number of outliers known*

Consider next the situation where the total number of outliers $m+r=k$ is known, but it is not known how many of these are on each side of the data. The most common statistic then seems to be the Tietjen – Moore E_k^* statistic defined as follows:

(i) Rank the data in ascending order of $|X_i - \bar{X}|$. Then using this ordering;

(ii) Compute

$$\bar{X}_k = \sum_1^{n-k} X_i/(n-k)$$

$$S_k^* = \sum_1^{n-k} (X_i - \bar{X}_k)^2$$

then $E_k^* = S_k^*/S$ is the statistic for a total of k outliers.

As before, exact distributions of this quantity are not known, but fractiles have been simulated by Tietjen and Moore (1972).

Actually, the E_k^* statistic has a serious defect, as pointed out in Hawkins (1978b). To see this, take the hypothetical case

$$X_1, X_2, \ldots, X_{n-2} \sim \mathrm{N}(0, 1)$$

$$X_{n-1} = 10$$

and suppose for ease of notation that X_1 is the smallest of $X_1, \ldots, X_{n-2}$. Faced with this sample and the information that a single outlier is present, one will surely agree that X_{n-1} is the outlier. Suppose, however, that we now add an observation

$$X_n = \delta \qquad \delta > 10$$

If we are told that two outliers are present, then it can easily be verified that the Tietjen – Moore statistic will test X_n and X_1 if δ is large, but X_{n-1} and X_n if δ is small. The prospect of rejecting X_1 as an outlier in preference to X_{n-1} seems totally unreasonable, as does the dependence of which of them is tested on the magnitude of X_n, and so constitutes an unhappy property of the statistic as originally defined.

This defect may be avoided by means of a slight modification of the test statistic due to Rosner (1975):

(i) Select as X_n that observation maximizing $|X_i - \bar{X}|$;

(ii) Recompute $\bar{X}_n$ on the sample remaining after deleting this X_n;

(iii) Repeat steps (i) and (ii) using the reduced sample, and continue until k of the X's have been deleted.

This modification ensures that the procedure satisfies the reasonable hierarchic requirement that the addition of a new observation, more extreme than any in the sample, should not affect one's assessment of the rankings of the aberrations of the remaining data.

The test statistic derived in this way cannot exceed E_k^*, since the method of construction ensures that the sum of squared deviations of the observations retained is at least as small as S_k^*. Simulated fractiles of this statistic (which we will denote E_k) are given in Hawkins (1978b), and are listed in Appendix 3. We may also note that these fractiles may be used to provide conservative fractiles for the $S_{m,r}/S$ statistics, since by definition such a statistic must exceed E_k; however, this bound may be very conservative, as one may verify simply by comparing the fractiles of L_k with those of E_k. The situation in which this great conservatism will arise is when m and r are very different.

We should note that it is important when using E_k that the value specified for k be correct. If there are fewer than k outliers, then these outliers can nevertheless so inflate S that E_k is significant, thus leading to the rejection of more outliers than are actually present, while if there are more than k outliers, then the masking effect may make E_k non-significant, leaving one to conclude wrongly that outliers are not present. This point is brought out *interalia* by studies of the power of Grubb's double-outlier statistic (Hawkins, 1978a) and by the influence curves given in Prescott (1978). These influence curves are contours of a given outlier statistic as a function of a set number of observations when the remainder of the 'sample' are set equal to expected normal scores. The curves provide an excellent qualitative basis for comparison of the likely performance of different outlier statistics under different models of the number and location of the outliers present. In particular, Prescott shows that the Tietjen–Moore statistic L_3 for three outliers is nearly as powerful against two outliers as is L_2, the statistic for exactly two outliers. Thus the E_k family are not robust against a mis-specification in either direction of the actual number of outliers in the data, and hence should not be used naively in the common situation in which the value of k is not completely certain. On the brighter side, this lack of robustness can be turned to good advantage. If one merely wishes to pre-screen a set of data for outliers without actually deciding how many are present, then one can simply apply the E statistic with k a slight overestimate of the number of outliers that one actually anticipates. The high power for a number of outliers lower than this value implies that the screen will be quite effective.

5.1.3 *Number of outliers unknown*

Next in order of difficulty comes the situation in which not even $m + r$ is known. This situation presents considerably more difficulties than do the cases considered in the two previous sections.

Basically, the approaches to this situation divide into two types: in the first, some auxiliary rule is used to estimate m and r, then a statistic suitable for that m, r pair used. As a variant of this approach, one might use a statistic such as E_k or E_k^* in conjunction with a rule for estimating $k = m + r$. The other approach uses the statistics themselves in some joint or stepwise way to decide on both the presence and number of outliers.

The superficially most appealing auxiliary rules for estimating m and r seem to be those based on gaps. Tietjen and Moore (1972), for the one-sided statistic L_k, suggest the rule: select as k that value for which the gap $X_{(n-k+1)} - X_{(n-k)}$ is a maximum. Their simulations show that if, after selecting k in this way, one uses the L_k statistic at a nominal significance level of 0.025, an overall size of about 0.05 results.

A refinement of this rule is proposed by Tiku (1974) who takes note of the fact that the gaps nearer the edge of the sample have larger expectation than those near the centre. His auxiliary rule uses the normalized gap $(X_{(n-k+1)} - X_{(n-k)})/(\mu_{n:n-k+1} - \mu_{n:n-k})$ where $\mu_{n:i}$ denotes the expectation of the ith order statistic of a sample of size n from an N(0, 1) distribution. Tiku, however, does not give any indication of how to adjust the nominal significance level of the outlier test specific to this k to achieve the size desired for the test.

Two points should be noted about these rules. The first is that the Tietjen–Moore rule will always test for a smaller number of outliers than the Tiku rule. The second point may be illustrated by the following simple example. Suppose that

$$X_1, X_2, \ldots, X_{n-2} \sim \mathrm{N}(0, 1)$$
$$X_{n-1} = 10$$
$$X_n = d > 10$$

Clearly if d is near 10, both rules will lead one to test for two outliers; however, if $d \gg 10$, then both tests will lead to a test for a single outlier. However, it is clear from the compactness of the set of $n - 2$ inliers that two outliers are present. This suggests that these gap-based rules are ineffective if the outliers come from markedly different distributions, which is precisely the situation in which one would use L_k or E_k in preference to, say, the Murphy statistic.

The possibility of using a rule which looks at the values of a test statistic jointly for several values of k suggests the following procedure when outliers are believed to come from the same distribution.

Using Murphy's (1951) optimality results for several outliers from the same distribution, compute for each k

$$B_k = c_k \sum_{i=1}^{k} (X_{(n-i+1)} - \bar{X})/\sqrt{S} \tag{5.6}$$

where the c_k are norming constants designed to make the B_k comparable for different k.
To find the c_k, let

$$\bar{X}_k = \sum_{i=1}^{k} X_{(n-i+1)}/k$$

$$\bar{X}'_k = \sum_{i=1}^{n-k} X_{(i)}/(n-k)$$

$$S_k = S - k(n-k)(\bar{X}_k - \bar{X}'_k)^2/n$$

Then

$$t_k = \{k(n-k)/n\}^{1/2}(\bar{X}_k - \bar{X}'_k)/\sqrt{S_k} \tag{5.7}$$

is the conventional two-sample t-statistic for comparing the 'samples' $X_{(1)}, \ldots, X_{(n-k)}$ and $X_{(n-k+1)}, \ldots, X_{(n)}$.

Some simple algebra then shows that

$$\sum_{1}^{k} (X_{(n-i+1)} - \bar{X}) = k(n-k)(\bar{X}_k - \bar{X}'_k)/n$$

and

$$B_k = c_k\{k(n-k)/n\}^{1/2} t_k(1 + t_k^2)^{-(1/2)} \tag{5.8}$$

Now if t_k is computed for an arbitrary ordering of the X_i, then clearly it follows a scaled t-distribution with $n + v - 2$ degrees of freedom for all values of k. Hence $c_k = \{k(n-k)/n\}^{-(1/2)}$.

The procedure then is: compute B_k (or equivalently t_k) for all k in the reasonable range anticipated *a priori*, and use the largest of these values to indicate both whether outliers are present and how many, and which observations they are.

If one has no prior idea of either how many outliers are to be anticipated or on which side they occur, then the range of k is 1, 2, ..., $n - 1$. Of course, since it is specified that all outliers have the same distribution, the conclusion of $k > n/2$ outliers on the right is logically equivalent to the conclusion of $n - k$ outliers on the left.

For the range $k = 1, 2, \ldots, n-1$, the statistic $\max|t_k|$ is well known. It arises both in automatic interaction detection (Kass, 1975; Scott and Knott, 1976) and in cluster analysis (Hartigan, 1975). Exact distributional results are known for the case σ known ($\nu \to \infty$), n small (Kass, 1975) and asymptotic results for the case $n \to \infty$ (Scott and Knott, 1976). The latter results are of especial interest. They show that, under H_0, the maximum value of $|t_k|$ occurs for $k = n/2 + o(\sqrt{n})$. (Actually, a proof of this result seems not to have been published – it is attributed to a personal communication from L. J. Savage. However, it is well supported by empirical evidence.)

The value of $\max(n+\nu-2)^{1/2}\,|t_k|$ can be shown to be of order $0.8\sqrt{n} + o(1)$ under H_0.

These two results are very interesting. They show a tendency, if H_0 is true but rejected, to divide the sample nearly in half. Furthermore, suppose the sample consists of $n-k$ inliers $N(0, 1)$, and k outliers $N(d, 1)$ say. Then if n and k are both moderately large

$$(n+\nu-2)^{1/2}|t_k| \approx \{k(n-k)/n\}^{1/2}(d-0) \tag{5.9}$$

If k is fixed while $n \to \infty$, this clearly tends to d, and so the probability of a correct identification goes to zero, since asymptotically $d < 0.8\sqrt{n}$. More realistically, if $r = k/n$ is fixed, then

$$(n+\nu-2)^{1/2}|t_k| \approx \{r(1-r)\}^{1/2}d\sqrt{n} \tag{5.10}$$

and clearly for the probability of a correct identification to be high asymptotically it requires $\{r(1-r)\}^{1/2}d > 0.8$.

If a binomial model of contamination is used with a fixed probability of r, then the following results show how large d must be for each r for this condition to be met.

r	d
0.01	8.0
0.05	3.7
0.10	2.7
0.15	2.2
0.20	2.0
0.50	1.6

Interestingly, the small values of r (which one expects to be the range of values of interest) require relatively very large values of d for the outliers to be detected with acceptably high power.

Despite the broad generality of this method, these minimum d

values are rather high, and clearly some modification to the test is needed to produce acceptable performance for moderate r and d. One possible modification would be to restrict the range of k. With the binomial model, this can be done quite easily and sensibly provided the probability of contamination is known, at least approximately. It is merely necessary to choose as the range of k those values having appreciable probability. This having been done, one might obtain fractiles of max $|t_k|$ by simulation.

Generalizing this approach slightly to include the E_k and L_k statistics, consider any statistic T_k designed to test for the presence of k outliers and, for a given sample, let a_k be its significance under H_0, and let w_k be a suitably chosen weight. Consider the procedure which concludes that there are k outliers if

$$\begin{aligned} w_k a_k &< w_j a_j \qquad j \neq k \\ w_k a_k &< a_0 \end{aligned} \tag{5.11}$$

The range of j may be taken to be $1, 2, \ldots, n-1$, and w_J set to infinity if we do not wish to test for J outliers.

Two unresolved elements in this procedure are the choice of the relative magnitude of the weights w, and the choice of a_0 to ensure a test of the required significance level. The second of these issues may be resolved by noting that the critical region is the union of the events $a_k < a_0/w_k$, and so has a size which, by Bonferroni's inequality, does not exceed $a_0 \Sigma w_k^{-1}$. Thus given the weights w, if one chooses $a_0 = \alpha/\Sigma w_k^{-1}$, an overall significance level not exceeding α is guaranteed, though this bound will be highly conservative if (as is the case with t_k, E_k and L_k) the statistics for different k are strongly positively correlated with one another.

The first issue is much less clear-cut. One attractive possibility is to make w inversely proportional to the *a priori* probability of k outliers, if this is known. However, it is unlikely that the procedure is too sensitive to the precise values chosen.

It may be noted that for all three of t_k, E_k and L_k this procedure is very susceptible to declaring the wrong number of outliers (see also Section 5.3). To illustrate the problem, consider the situation with t_k. Suppose one has 10 data values, seven of which are the normal scores for a sample of size 7 from N(0, 1) while the remaining 3 are normal scores from N(δ, 1). Provided $\delta > 2.2$, the order statistics of the 'sample' are -1.35, -0.76, -0.35, 0, 0.35, 0.76, 1.35, $\delta - 0.85$, δ, $\delta + 0.85$. The following are the values of t_k and 10000 times a_k for $k = 2$ to 6 as a function of δ.

	k				
δ	2	3	4	5	6
2.2	9.7	10.2	11.7	12.5	10.2
	92	70	31	20	70
2.4	11.1	12.1	13.4	13.9	11.2
	43	25	13	10	41
2.6	12.6	14.2	15.2	15.4	12.2
	19	8	5	4	24
2.8	14.2	16.5	17.0	16.9	13.3
	8	2	2	2	13
3.0	15.9	18.9	19.0	18.5	14.5
	3	1	1	1	7
3.2	17.8	21.5	21.1	20.2	15.7
	1	0	0	0	4

It is of great interest that when δ is relatively small, the a_k favour neither 3 nor 4 outliers, but 5, and it is only when δ is in excess of 3.0 that the correct number of outliers is indicated in any unequivocal sort of way, and even here the values of t_4 and t_5 are of fairly similar magnitude to t_3.

Thus the use of t_k can easily lead to identification of spurious outliers in this procedure. The use of L_k and E_k, by contrast, can lead to the underestimation of the number of outliers provided the outliers are well separated from the main body of the data. Two effects lead to this; one is that if k outliers are actually present (and may be assumed to be the k most extreme values of the sample) then the observations retained after their deletion constitute a random sample of size $n-k$ from $N(\xi, \sigma^2)$. However, the distribution theory of L_k and E_k assumes them to be the inner $n-k$ order statistics of a sample of size n. As even the most cursory glance at a table of expected order statistics soon shows, using the latter assumption would imply a much larger variance than σ^2, and so there is significant loss of power for the tests for k, $k+1, \ldots$ outliers. The second effect potentially is the more serious. Without going into great detail, it is that if, for example, there are two outliers displaced respectively by amounts $c\delta$ and δ ($\delta > 0, c > 1$) then for large c, the values of E_1 and L_1 attain an arbitrarily larger significance than do E_2 and L_2 as $\delta \to \infty$. This procedure is thus incapable of locating important outliers when they are overshadowed by other more aberrant outliers.

These defects in the general procedure are serious enough to relegate it to second place behind the more successful stepwise procedures discussed in the following section.

5.2 Stepwise procedures

The alternative approach is the use of a stepwise procedure. The original variant of this approach is to test the sample $X_1, \ldots, X_n$ for a single outlier. If the outlier test is significant, then the test is repeated on the subsample remaining after deletion of the outlier. This process is continued until the single-outlier statistic becomes non-significant. The subsample remaining at that point is accepted as containing no further outliers. We will term this the 'forward selection' approach, because of the connection with the forward selection procedure of multiple regression, as set out in Chapter 8. We will denote by $T_{n:i}$ a generic test statistic applied to the ith most outlying observation in a sample of size n.

The details of the behaviour of the forward selection method under the null hypothesis are quite well understood when the optimal statistic for a single outlier $T_{n:1} = (X_{(n)} - \bar{X})/\sqrt{S}$ is used. First, if the α fractile of $T_{n-i+1:1}$ is used at the ith stage, then the probability of declaring at least 1 outlier if there are none is α.

Suppose next that there is a single outlier. Then the probability of declaring more than one outlier may be reasoned out as follows: conditionally on the significance of the first stage, the probability that it was the real outlier that was detected is near 1 whenever the outlier is markedly aberrant (see, for example Chapter 3). If the correct outlier has been rejected, then what remains is a legitimate sample of size $n-1$, and so the probability of a further significance is again α.

Now since the process of declaring two outliers requires significance at the first stage, those situations in which the outlier is not markedly aberrant have low enough power not to affect the two-stage probability perceptibly.

This reasoning may be taken further and leads to the conclusion that regardless of the real number of outliers, the probability of finding one spurious outlier is of order α, that of finding two is of order α^2, and so on.

This situation is entirely satisfactory – the probability of type I error is well under control, and is well understood. However, the situation as regards type II error is not at all satisfactory and for this reason the forward selection method may behave poorly. The problem is the so-called 'masking effect', and was discussed clearly in the paper of Pearson and Chandra Sekar (1936). An illustration of the effect was given in Section 5.1.

Clearly, this effect is not to be found only when the two outliers are very similar in value (though this is when it is most severe) and so one can infer that the primary damage to the power results from the inflating effect the additional outlier has on the variance S. This makes it clear that the critical effect of the power dropping to zero is to be expected when the underlying variance is not estimated well by S – that is, when ν is small, or zero, and n is not large. In fact, if σ is known, then the effect of the second outlier is to alter $T_{n:1}$ from $(n-1)\delta/n$ to either δ (if the second outlier is on the opposite side of the sample) or $(n-2)\delta/n$ (if it is on the same side) with effects on the power which are clearly minor, even for n fairly small.

Some figures on the magnitude of the masking effect are given in Hawkins (1973), and these suggest that, if two outliers are actually present, the masking effect is not of serious scale if very roughly $n+\nu>20$ and $n>5$, and so double outliers may be detected fairly reliably in samples which are not excessively large, especially if external information on σ is available. This fact provides only partial consolation, since in large samples, the possibility of three or more contaminants becomes increasingly probable and with it, a return to the masking problem.

TABLE 5.1 $n=15, \nu=0, \alpha=0.05$

	τ				
δ	1	3	5	7	9
1	0.02				
3	0.36	0.14			
5	0.89	0.67	0.26		
7	1.0	0.98	0.80	0.35	
9	1.0	1.0	0.99	0.86	0.42

$n=15, \nu=5, \alpha=0.05$

	τ				
δ	1	3	5	7	9
1	0.03				
3	0.41	0.20			
5	0.93	0.80	0.46		
7	1.0	0.99	0.94	0.68	
9	1.0	1.0	1.0	0.98	0.82

Table 5.1 gives some illustrative figures on the masking effect. In it, a sample of size n is studied which contains $n-2$ N(0, 1) values, a N(δ, 1) and a N(τ, 1). The table shows the probability of detecting at least one of these two contaminants by forward selection.

The figures in the table illustrate the heuristic rule about the sample size above which masking becomes a relatively minor problem. Note also that the effect is most severe if $\delta = \tau$, as was noted earlier.

A variant of the stepwise procedure which is immune to the masking effect is provided by the following 'backward elimination' procedure: set an upper limit K on the number of outliers to be tested for. Remove the K most extreme observations. Then, at each stage, test the most inlying of the suspected outliers for compatibility with the 'clean' observations. If it is compatible, then include it with the clean observations and repeat the procedure with the next suspected outlier. Otherwise stop, and conclude that that observation, together with those more extreme, is an outlier.

Note that, provided the value of K specified does not exceed the actual number of outliers, then this procedure cannot be affected by masking since, at each stage, only the original $n-K$ clean observations and those suspected outliers that have passed scrutiny are used as the basis for comparison of the next suspected outlier.

While this method is acceptable from the point of view of masking, it provides some difficulties as regards type I error, and the distributional problems associated with finding suitable fractiles have hardly been touched in published work. The principle behind the sequence of tests is quite straightforward. It is analogous to the problem of deciding by what order polynomial to de-trend a time series (and the excellent account of this problem given in Anderson (1971) is especially appropriate). In both problems, masking effects create the danger of stopping too soon in a forward selection procedure, and so make it advisable to proceed by backward elimination.

The backward elimination procedure consists of setting up the K test statistics $T_{n:1}, \ldots, T_{n:K}$, each corresponding to the testing of the ith suspected outlier against the $n-i$ observations more inlying than itself. These are then tested in reverse order, starting with $T_{n:K}$. The ith is tested at significance level α_i and if the test is significant, the ith suspected outlier is rejected together with all observations that are more outlying.

However, despite the otherwise close analogy with the problem of deciding the correct order of polynomial trend in a time series, the backward elimination procedure does not provide a statistically similar test. To see this, note that the problem consists of a multiple decision between the hypotheses

$$
\begin{aligned}
&H_0 : \text{No outliers present}\\
&H_1 : \text{Exactly one outlier present}\\
&\quad\vdots\\
&H_k : \text{Exactly } k \text{ outliers present}\\
&\quad\vdots\\
&H_K : \text{Exactly } K \text{ outliers present}
\end{aligned}
$$

Suppose that in fact k outliers are present, and consider the test of H_i for $i > k$. The distribution of the test statistic (unlike in the polynomial regression case) depends on the actual value of k. Ignoring type III errors, which are both negligible in probability and inconsequential in effect, if k outliers are present, then the statistic $T_{n:i}$ is actually that corresponding to the testing of the $(i-k)$th observation in a sample of size $n-k$. That is, we test the value of $T_{n:i}$ against the null fractile of $T_{n-k:i-k}$. Thus the null distribution of $T_{n:i}$ depends on the true value of k, which is not specified by the hypothesis null to H_i, namely $H_0 \cup H_1 \cup \ldots \cup H_{i-1}$. This implies that any static test procedure will not provide a similar critical region. In particular, testing $T_{n:i}$ against H_0 fractiles of $T_{n:i}$ may lead to a significantly wrong probability of type I error if there are in fact any outliers present.

It is suggested by heuristic reasoning, and has been verified by simulation, that the most stringent test of H_i is provided by assuming as null hypothesis that $i-1$ outliers are present. Thus a test whose size is uniformly $\leqslant \alpha$ is provided by testing $T_{n:i}$ against the α_i fractile of a $T_{n:i}$ statistic based on a sample size of $n-i+1$ with $\Sigma\alpha_i = \alpha$. This procedure has the great advantage of only using standard tabled fractiles of $T_{n-1+1:1}$ (which in turn are approximated with excellent accuracy by the t-distribution), but turns out, as will be shown below, to be very conservative.

It may be noted, as a matter of interest, that the usual analysis-of-variance decomposition for the removal of successive outliers shows that

$$E_k = \prod_{i=1}^{k} \{1 - (n-i+1)/(n-i)T_{n:i}^2\} \tag{5.12}$$

a result which serves to tie in the stepwise approach with the k-outlier statistic E_k. A similar relationship exists between L_k and the sequence of one-sided outlier test statistics.

It will be shown in Chapter 7 that, in terms of a multiple-regression formulation of the multiple outlier problem, the $T_{n:i}$ are equivalent to successive partial correlations in a multiple regression, and E_k to the multiple correlation coefficient of the overall regression. In regression, both the partial and the multiple correlations have their uses – it is well known that there can exist highly significant multiple correlations even where there are no significant partial correlations, and vice versa. In the same way, one can envisage situations in which the use of the statistic E_k would not lead to the same conclusion as the stepwise sequence based on the $T_{n:i}$, and neither statistic dominates the other in performance.

5.3 Performance of the procedures

Relatively little is known about the performance of any of these procedures in the presence of multiple outliers. Tietjen and Moore (1972) present simulation data on the performance of their statistic L_k, supplemented with a gap rule for selecting a suitable k. Hawkins (1978a) presents analytic results on the relative performance of L_2, $T_{n:2}$ and the Murphy statistic $(X_{(n)} + X_{(n-1)} - 2\bar{X})/\sqrt{S}$ for detecting two outliers. However, neither paper does more than illustrate the acceptable performance of the procedures with two well-defined outliers.

Rosner (1975) presents a comparison of seven double-outlier test cases, all for $n = 20$, $\nu = 0$, of several statistics – the backward elimination use of $T_{n:i}$, of the studentized range, and of his R statistic (defined below); the kurtosis, and the Tietjen–Moore double-outlier statistic E_2^*. Despite the rather restricted range of situations simulated, it emerged fairly clearly that for $n = 20$ the performance of all procedures except the studentized range was quite acceptable.

A much more serious problem is the choice between forward selection and backward elimination if it is decided to use a stepwise procedure. We note that both procedures use the identical sequence $T_{n:i}$, but test them in a different order. While in the forward selection procedure, $T_{n:i}$ is tested at a significance level α, in the background elimination procedure, it is tested at a significance level α_i, where $\alpha_i < \alpha$. This latter test is clearly more stringent.

One would hypothesize that forward selection would perform

acceptably when the number of outliers is small, n is large and there is appreciable external information on σ. It is also likely to be satisfactory if there are several outliers, but each is much more outlying than those following it. For example, if there are several outliers, but the ith has a relative displacement of say 1.5^i, then masking is likely not to be a severe problem, and forward selection may perform quite well.

If on the other hand, the outliers are all of more or less equal size, then a possible competitor to backward elimination would be the approach using the automatic interaction detection (AID) statistic.

In order to obtain some feeling for the relative performance of the three approaches of forward selection, backward elimination and AID, a small simulation exercise involving 200 samples was carried out. In it, it was specified that at most three outliers were present, and three contaminants were in fact introduced by adding on contaminating amounts δ_1, δ_2 and δ_3 to X_1, X_2 and X_3. Three situations were simulated – $n = 10$, $\nu = 0$; $n = 10$, $\nu = 10$; and $n = 20$, $\nu = 10$. In all, $\alpha = 0.1$ was used.

The forward selection fractiles were chosen as the $0.1/2(n-k)$ fractiles of the corresponding t-distributions, and the backward elimination fractiles, following on earlier comments, as the $0.1/6(n-k)$ fractiles of the same t-distributions. The AID fractiles were estimated by a separate simulation of 2000 samples.

The results are summarized in Table 5.2, the rows labelled FS, BE and AID. The table shows the percentage of tests yielding 1, 2 or 3 significant outliers ($\geqslant 3$ in the case of AID). It is of interest to note that even for $n = 20$, $\nu = 10$, when one might have thought that there was very adequate information on σ, there was enough evidence of masking when all δ_i were equal to 4 to make one rather apprehensive about using forward selection. On the other hand, if the δ_i are not equal, forward selection is quite effective.

Backward elimination performs quite well, even though a separate simulation of the null case showed that the test is very conservative in declaring two or more outliers. This conservatism shows up the desirability of some modification which would retain control over type I error, but be less conservative.

AID performed well when all δ_i were equal. When the δ_i were greatly unequal though, it was less effective. Perhaps this defect could be overcome by a repeated application of the test; eliminating a block of outliers and then repeating the test on the remaining observations.

TABLE 5.2 Powers of the procedures

n, v	10,0			10,10			20,10		
number found	1	2	$\geqslant 3$	1	2	$\geqslant 3$	1	2	$\geqslant 3$
$\delta = 1, 3, 5$									
FS	31	8	2	40	34	6	48	38	4
BE	14	11	2	45	28	2	51	28	3
AID	0	10	4	12	48	20	4	22	18
ROS BE	20	16	6	41	55	3	35	60	2
ROS FS	17	4	12	51	36	8	50	36	8
$\delta = 2, 4, 6$									
FS	25	14	1	25	52	14	23	60	14
BE	12	22	4	33	46	8	36	50	20
AID	0	20	6	8	54	33	2	36	34
ROS BE	26	14	4	15	72	14	14	82	5
ROS FS	14	9	8	31	43	25	28	60	11
$\delta = 4, 4, 4$									
FS	2	0	0	15	7	11	15	18	30
BE	0	1	19	8	8	34	12	17	39
AID	0	1	42	0	5	92	0	2	83
ROS BE	11	9	6	18	54	18	16	66	17
ROS FS	4	2	10	20	24	44	26	28	35

The study also illustrates AID's propensity for finding more outliers than are actually present. The percentage of the samples giving four or more significant outliers on the AID criterion is shown in Table 5.3. While both the forward selection and the backward elimination methods control the probability of declaring any spurious outliers, we see that with the AID test statistic this probability may be alarmingly high. This alone tends to militate against the use of the AID statistic.

The forward selection criterion has an appeal that is difficult to ignore in that its fractiles are easy to compute (being simply the fractile for the most extreme outlier in sample of size $n - k + 1$). Thus one wonders whether the forward selection procedure, applied to a test statistic which is not sensitive to masking, could be used to good effect.

TABLE 5.3 4 or more outliers found by AID

n, v	10, 0	10, 10	20, 10
$\delta = 1, 3, 5$	1	11	12
$\delta = 2, 4, 6$	0	6	18
$\delta = 4, 4, 4$	6	21	36

One such test statistic is that due to Rosner (1975). The test statistic applied to X_i is

$$R_i = (X_i - \hat{\xi})/\hat{\sigma}$$

where, if at most K outliers are anticipated,

$$\begin{aligned} \hat{\xi} &= \sum_{K+1}^{n-K} X_{(j)}/(n-2K) \\ \hat{\sigma}^2 &= \sum_{K+1}^{n-K} (X_{(j)} - \hat{\xi})^2/(n-2K-1) \end{aligned} \tag{5.13}$$

i.e. $\hat{\xi}$ and $\hat{\sigma}$ are the symmetrically trimmed mean and standard deviation of the data. If external information U on σ is available, one would utilize it by adding U to the numerator, and v to the denominator of $\hat{\sigma}^2$.

Since at most K outliers are anticipated, the estimators of location and scale are clearly not directly contaminated (though if outliers are present, they do suffer from a more subtle form of contamination discussed below).

Rosner originally proposed the statistics R_i in conjunction, not with forward selection, but with backward elimination. In this latter form, one would expect it to be less powerful than the optimal procedure using $T_{n:i}$, but possibly desirable on the grounds of its computational simplicity, only a single pair of location/scale statistics being computed for all K outlier statistics. It is in fact mainly on these grounds that Rosner proposed the statistics.

The rows labelled 'ROS FS', and 'ROS BE' in Table 5.2 show the performance of this test statistic in its forward selection and backward elimination versions. One should note that in the original definition, Rosner defined the critical regions C_i for R_i by

$$\Pr[R_i \in C_i | H_0] = \beta \qquad i = 1, 2, \ldots, K$$

$$\Pr[\bigcup_{1}^{K} (R_i \in C_i) | H_0] = \alpha$$

As already noted, this definition is a poor one, since when outliers are present, it does not adequately control the probability of declaring more outliers than are actually present. Nevertheless, for consistency with existing literature, this definition was retained.

Comparison of these additional entries, both with one another and with the earlier ones, proves to be very informative. Taking as a rough summary statistic the overall percentage of all outliers located by each of the procedures gives the following results:

FS	35	ROS FS	45
BE	38	ROS BE	47
AID	53		

The most effective technique is AID, but both it and ROS BE are, as noted, suspect for size reasons. ROS FS performs very well – noticeably better than FS, and probably better than ROS BE, bearing in mind the latter's size problems.

Overall, one is left with the impression that, unless one can solve the size difficulties of BE, ROS FS is most effective for general use, though it has the severe disadvantage that there are no extensive tables of or algorithms for its fractiles.

Fortunately, it is possible to remove almost entirely the objection of uncertain size of the BE procedures. The price one pays for this is that it is no longer possible to use only standard tabled functions; in fact it is necessary to resort to simulation. We start off with the proposition that if k outliers are present and are more extreme than the inliers then the probability of declaring more than k outliers should equal α for all k. From this proposition, we proceed as follows.

Consider first $T_{n:K}$. We require that this be significant with probability α when in fact $K-1$ outliers are present; thus the appropriate fractile $C_{n:K}$ is the α fractile of the $T_{n-K+1:1}$ statistic and may be found from the theory discussed earlier for a single outlier.

Suppose now that $C_{n:k+1}$ up to $C_{n:K}$ have been chosen. To find $C_{n:k}$ note that our requirement is that given $k-1$ outliers,

$$\Pr[\bigcup_{i=k}^{K} (T_{n:i} > C_{n:i})] = \alpha \tag{5.14}$$

Since $C_{n:k+1}$ up to $C_{n:K}$ have been determined in earlier steps, this gives us an equation in the single unknown $C_{n:k}$, which we then solve. Proceeding backwards, one ultimately solves for $C_{n:1}$, after which the entire procedure is fully specified. The major difficulty that arises in implementing this procedure is that there is no known analytic form for the joint cumulative distribution function, and it is thus necessary to use simulation. The approach can be partly analytic, since, letting A_i be the event $T_{n:i} > C_{n:i}$, we have

$$\Pr\left[\bigcup_{i=k}^{K} A_i\right] = \Pr[A_k] + \Pr\left[\bigcup_{k+1}^{K} A_i\right] + \Pr\left[A_k \cap \bigcup_{k+1}^{K} A_i\right] \tag{5.15}$$

The last term in the expansion is clearly negligible in comparison with the second term. Ignoring it, we get

$$\Pr[A_k] \approx \alpha - \Pr\left[\bigcup_{k+1}^{K} A_i\right]$$

From $\Pr[A_k]$ it is an easy matter to deduce $C_{n:k}$ since this is simply a fractile of the distribution of $T_{n-k+1:1}$, so using simulation to estimate $\Pr\left[\bigcup_{k+1}^{K} A_i\right]$, one may use the known theory of the distribution of the single-outlier statistic in a sample of size $n - K + 1$ to find $C_{n:k}$.

This part-simulation, part-analytic approach seems likely to yield more accurate estimates of the $C_{n:k}$ than would a pure simulation approach; this quite reasonable expectation was borne out by a simulation study in which the method was found to have considerably smaller sampling variance than pure simulation approach, and to display no evidence of bias. A table of the resulting fractiles of $T_{n:i}$ is given in Appendix 4.

In computing these fractiles, it was noted that except for $n = 10$, $\nu = 0$ or 5, the term $\Pr\left[\bigcup_{k+1}^{K} A_i\right]$ did not exceed $\alpha/4$. Thus except for n and ν both small, it seems that little error would be made by testing $T_{n:i}$ against fractiles of $T_{n-i+1:1}$, and hence the tests applied for $k + 1, \ldots. K$ outliers come almost free of charge. This fact is initially somewhat surprising, but means no more than that the inliers of a normal sample constitute a much more platykurtic set than does a random sample. Thus if one is willing to accept a test whose actual size may be 25 per cent in excess of the nominal significance level, then one may ignore the values in Appendix 4, and use the values of Appendix 1, or the Bonferroni inequality which provides such an excellent approximation to the values in Appendix 1.

The performance of this final version of the backward elimination procedure is set out in Table 5.4. As the table makes clear, its

TABLE 5.4 Power of final version of BE

n, ν	10, 0			10, 10			20, 10		
number found	1	2	3	1	2	3	1	2	3
$\delta = 1, 3, 5$	23	17	10	46	34	7	49	36	8
$\delta = 2, 4, 6$	14	18	10	25	46	24	25	56	17
$\delta = 4, 4, 4$	1	1	38	10	13	56	16	14	58

performance is very good. Overall, it located 51 per cent of all outliers in the samples, and hence was noticeably more effective than the other methods. It is especially significant that its performance was relatively poor only when one of the δ_i was 1 – a situation in which high power is not to be expected.

We may note that the procedure used here is a special case of a more general class in which

$$\Pr\left[\bigcup_{k}^{K} (T_{n:i} > C_{n:i})\right] \leqslant \alpha$$

Amongst members of this class, it has the property of applying the least stringent test to the largest number of outliers. This property, while it leads to an optimal procedure if there are, in fact, K outliers, is non-optimal if K is some generous overestimate chosen to provide protection against the possibility of a greater number of outliers than is really anticipated. If this is the case, then one would do better to allocate a size strictly less than α to these very large k values, and hence be able to use a less stringent test for smaller k. However, as we have already seen, unless both n and ν are small, the benefit one can achieve by doing this is very limited.

CHAPTER 6

Non-parametric tests

The corpus of theory on non-parametric approaches to outlier testing is not large, and given the strongly parametric nature of the outlier model, this is really not surprising. Most of the results that there are relate to slippage rather than outlier problems. Suppose that we are given X_{ij}, $j = 1$ to n_i, $i = 1$ to m, $n = \Sigma n_i$ observations from m groups. Let X_i denote an arbitrary observation from population i. The null hypothesis is

$$H_0 : \Pr[X_i > X_j] = \tfrac{1}{2} \qquad \text{all } i, j$$

A 'slippage to the right' model is that for some unknown i

$$\begin{aligned} H_1 : \Pr[X_i > X_j] &> \tfrac{1}{2} \qquad j \neq i \\ \Pr[X_j > X_r] &= \tfrac{1}{2} \qquad j, r \neq i \end{aligned}$$

and a two-sided equivalent would be that for some unknown i,

$$\begin{aligned} H_2 : \Pr[X_i > X_j] &\neq \tfrac{1}{2} \qquad j \neq i \\ \Pr[X_j > X_r] &= \tfrac{1}{2} \qquad j, r \neq i \end{aligned}$$

As the preceding chapters have made clear, there is no difficulty of principle involved in setting up a test for a single slipped treatment: one merely takes a convenient two-sample statistic, and applies it to the problem of comparing the ith observation or treatment with the aggregate of the remainder of the sample.

In the realm of non-parametric statistics, there are fewer optimality properties to guide one than there are in parametric statistics, and thus there is a wider range of possible test statistics, none of which dominates all others. Despite this fact, only two non-parametric slippage tests seem to be in common use: one due to Mosteller (1948). and the other to Doornbos (1966). Of the two, the former test is rather quicker to apply while the latter is likely to be more powerful. The evidence for this assertion will be examined later.

6.1 The Mosteller statistic

The test statistic proposed by Mosteller is defined as follows: find that treatment i containing the largest order statistic in the combined sample. The test statistic is the number of observations in this treatment exceeding in value all observations for every other treatment. If all $n_i = n$, Mosteller shows that the probability that his statistic is r is given by $n!\, m(nm - r)!/(mn)!(n - r)!$, and the resultant tables of fractiles are listed in Appendix 6. Mosteller and Tukey (1950) discuss the case when the n_i are not all equal, and provide some fractiles for that case.

The technique was proposed as a 'quick and dirty' method, and this author feels that, while it is certainly quick, it is dirtier than is necessary or appropriate. To justify this claim, consider the two-sample statistic on which it is based. The method would be given two samples $X_1, \ldots, X_n$ and $Y_1, \ldots, Y_m$, find N, the number of X_i exceeding all Y_j (assuming that the largest order statistic of the combined sample is an X). Compare this with the median test, which finds M, the median of the combined sample of X's and Y's, and then uses as test statistic N', the number of X_i exceeding this common median.

The statistic N seems inferior to N' in being much more strongly dependent on the value of the largest order statistic of the Y's, and this is surely a serious failing if the data follow a heavy-tailed distribution. Since it is for the heavy-tailed double exponential distribution that N' is the optimal two-sample test, this criticism is quite a serious one. Nevertheless, the Mosteller statistic has found its place in the literature, and has in fact been extended to the situation of multiple slipped treatments.

Joshi and Sathe (1978) propose an extension of Mosteller's test in which the treatments are ranked in order of their ith largest order statistics. The test statistic (which may be zero for $i > 1$) is the number of observations from the first-ranked treatment exceeding all values in all other treatments. For $i = 1$, this test is identical to the usual Mosteller test. Joshi and Sathe present some power calculations which show that the extended test can be more powerful than the standard test; however, the choice of a suitable i (this author has an intuitive liking for the median), and distributional properties for general i, both remain to be solved fully.

6.2 The Doornbos statistic

The specific procedure due to Doornbos uses as its test statistic the Wilcoxon statistic R_i defined as the sum of the ranks of the X_{ij} in the combined sample. While exact fractiles of the test statistic seem not to be known, the Bonferroni bound, as usual, is either exact or provides a very good approximation, and rejects H_0 if one of the R_i is significant at the α/m level.

A table of this Bonferroni approximation is given in Appendix 7.

Doornbos shows that this application of the Wilcoxon test leads to the same asymptotic relative efficiency relative to the Paulson statistic as the usual two-sample Wilcoxon test has relative to the t test. This efficiency is $3/\pi$ or 95 per cent.

The application of the test is quite straightforward. For slippage to the right one rejects H_0 on large values of $\max R_i$, while for slippage to the left, one rejects H_0 on small values of $\min R_i$. If the direction of slippage is unspecified, then one attaches to each R_i the two-sided significance of R_i. H_0 is then rejected if the smallest such α_i is sufficiently small.

The fractiles required are basically the α/m and $\alpha/2m$ fractiles of R_i based on sample sizes n_i and $n - n_i$, and the awkward values of the significance levels provide the only difficulty.

The normal approximation to the Wilcoxon fractiles, usually quite adequate even for small samples, has rather stringent sample-size requirements in the slippage test. This is because, for large m, one is interested in the extreme fractiles, where the normal approximation deteriorates rapidly. Nevertheless, the normal approximation has the desirable property of being slightly conservative, and so may reasonably be used outside the range of values covered in the table of Appendix 7.

6.3 Slippage in scale

The procedure just described is directed at situations in which the spread of the m groups is assumed constant. Another situation of interest is testing for slippage in a scale parameter. Provided all location parameters are equal, this can be tested with the Siegel–Tukey statistic (Lehmann, 1975) defined as follows: order the data and assign ranks of:

1 to the largest datum
2 to the smallest
3 to the second smallest
4 to the second largest
5 to the third largest

and so on. Then using these ranks, compute the Wilcoxon statistic.

The null hypothesis distribution of this statistic is clearly identical to that of the usual Wilcoxon statistic for location slippage. Small values of R_i correspond to large spread, and vice versa.

Suppose now that in one group, both the location and the spread have slipped upward. This can lead to a loss of power in both the location and the scale tests. The location test loses power since, as a result of the increased spread, the smallest order statistics of the slipped population may have very low ranks in the combined sample. This decreases the sum of ranks, and hence the power.

Similarly in the scale statistic, since large ranks and small ones are distributed amongst both the large and the small order statistics, any change in the location parameter of a population tends to move its mean rank towards $\frac{1}{2}(n + 1)$, and again there is a loss of power.

In the case of the scale test, it is tempting to try to correct the observations for any differences in location. One possible method of doing this is to subtract from each observation an estimate – such as the sample median – of the population's location parameter. This is acceptable, provided the sample sizes within each group are sufficiently large (Lehmann, 1975). No such correction for scale seems to have been applied to the location statistic.

6.4 Multiple slippage

Next, let us turn to the problem of testing whether k groups have slipped. The various possible situations are distinguished by knowledge or lack thereof of how many populations have slipped; how many of these have slipped to the left and how many to the right, and whether the populations have slipped to the same extent. Let us adopt the notation that k populations have slipped, r of them to the right. Non-parametric procedures can then be found by direct adaptation of the procedures discussed in Chapter 5 – the adaptation consists largely of replacing the parametric statistics by non-parametric equivalents. We will discuss the problem in terms of the Doornbos–

Wilcoxon test for location, but the discussion holds equally well for the scale problem using the Siegel–Tukey statistic.

Asymptotic theory may be derived in the following way. Assume all n_i equal to N, say. Then asymptotically

$$R_i \sim \mathrm{N}[\xi, \sigma^2] \tag{6.1}$$

where

$$\xi = N(Nm+1)/2$$
$$\sigma^2 = N^2(m-1)(Nm+1)/12$$

The R_i are clearly exchangeable variables, and sum to a constant. Thus their common intercorrelation is $-1/(m-1)$.

Defining $U_i = (R_i - \xi)/\sigma$, we then see that the $\{U_i\}$ are distributed like the quantities $(X_i - \bar{X})\{m/(m-1)\}^{1/2}$, where $X_1, \ldots, X_m$ constitute a random sample of size m from $\mathrm{N}(\xi, 1)$. Thus the distributions of statistics based on the ranked U_i (or equivalently on the R_i) may be derived by a direct adaptation of the theory relating to parametric models of $\mathrm{N}(\xi, \sigma^2)$ data in which ξ is unknown and σ known. This theory in turn follows from the general theory of Chapter 3 and 5 on letting v tend to infinity.

First, if it is believed that the slipped populations have all shifted by the same amount, then one would use statistics of the form $\Sigma_1^k R_{(i)}$ (or equivalently $\Sigma_1^k U_{(i)}$). Now clearly if k is known, then these are simply the asymptotic cases of the general Murphy statistics discussed in Section 5.1.1 If k is unknown, then one would think in terms of using the AID statistic, either in its standard form or in a variant form as set out in Section 5.1. Note that the fact that σ is known brings this situation directly into line with that studied by Kass (1975), and so his fractiles may be used for the conventional form of the AID statistic.

6.5 Amounts of slippage different

Next, let us consider the more likely situation in which the amount of slippage is not known, and presumed different for the different slipped populations. Recall that the statistic of Tietjen and Moore (1972) for k suspected slipped populations consists of $k+1$ group analysis of variance, in which one compares the k most extreme samples and a single group obtained by pooling all remaining inlying groups into a single sample. This suggests as an appropriate non-parametric test an analogous procedure using a non-parametric analysis of variance. One such is the Kruskal–Wallis statistic (see, for example, Lehmann,

1975). Let R_i denote the mean rank in the combined sample of the n_i observations in the ith group. Then the Kruskal–Wallis statistic is

$$W = 12\sum_1^m n_i(R_i - R)^2/n(n+1)$$

where the grand mean rank, R, is clearly $(n+1)/2$. This quantity is asymptotically distributed as χ^2_{m-1} as all n_i tend to infinity.

Now to apply this to the slippage problem, assume that all n_i are equal to N, and that k outliers are suspected. Arrange the groups so that $R_1, R_2, \ldots, R_k$ are k most extreme mean ranks. This ranking may be either one-sided or two-sided, depending on whether or not the slipped populations are specified *a priori* to have slipped in the same direction. Letting $\bar{R}$ denote the mean of these k mean ranks, it is easily shown that the Kruskal–Wallis statistic becomes

$$W_k = 12\left[\sum_1^k N(R_i - R_.)^2 + Nk^2(\bar{R} - R_.)^2/(m-k)^2\right]/n(n+1) \quad (6.2)$$

Provided N is large enough to justify the invocation of asymptotic normality, the distribution of W_k follows from limiting theory of the E_k statistic, for if E_k is computed on a sample of size m using ν external degrees of freedom as additional information on σ, then clearly W_k has the same distribution as

$$\lim_{\nu \to \infty} \nu[1 - E_k]$$

and provided N is at least moderately large, the asymptotic theory of E_k may be used to obtain the fractiles needed.

As an alternative to this limiting distribution, we may use either of two conservative upper bounds:

(i) W_k as computed is obviously less than the Kruskal–Wallis statistic computed using all m groups. Thus the upper α fractile of a χ^2_{m-1} distribution provides a conservative fractile for W_k. This bound is essentially a multiple comparison bound, and has the great advantage of being valid simultaneously for all k. Thus it is a legitimate bound whether k is fixed *a priori* or chosen after inspection of the data.

(ii) Ignoring the ranking of the R_i, W_k would follow a χ^2_k distribution. Thus an upper bound using Bonferroni's inequality assesses W_k against the α/M fractile of χ^2_k, where $M = m!/k!(m-k)!$.

Since the χ^2 statistic is, by its very nature, two-sided, both of these methods are legitimate for either one- or two-sided outlier tests. If k is specified *a priori*, then one has a choice of which of these two conservative fractiles to use, and one will actually use the smaller of the two.

The data in Table 6.1 illustrate the fractiles, and how conservative they actually are. The sizes quoted for the conservative fractiles were estimated from 2000 simulated samples of N(0, 1) variables. Next to each simulated fractile is listed the fractile obtained by extending the fractiles of E_k to $\nu = \infty$. Since this represents an extrapolation out of the range over which the fractiles of Appendix 3 were simulated, it can by no means be taken for granted that these fractiles are reliable; nevertheless, they agree well with the simulated fractiles except for $m = 30$, $k = 10$.

These figures show that the Bonferroni bound is fair for k small, but for larger k, it soon exceeds the multiple comparison bound. The latter bound is quite good if k is of order m. Of course, this result is actually quite disappointing. One of the advantages of a slippage test over the m-way Kruskal–Wallis test is its greater power. Obviously this advantage is lost if one uses the multiple comparison fractiles, and so the adequacy of the approximation merely suggests that the slippage test is not much more powerful if a large fraction of the populations have slipped.

TABLE 6.1

m	k	Bound (i)	Bound (ii)	Size	5% Fractile
5	1	9.49	6.63	0.05	6.3 (6.63)
	2	9.49	10.60	0.03	8.4 (8.80)
	3	9.49	12.84	0.05	9.6 (9.06)
10	1	16.92	7.88	0.06	8.3 (7.88)
	2	16.92	13.60	0.01	11.0 (11.56)
	3	16.92	18.11	0.01	13.2 (13.83)
	5	16.92	24.20	0.02	15.8 (15.71)
20	1	30.14	9.14	0.05	9.1 (9.14)
	2	30.14	16.49	0.01	13.8 (14.21)
	3	30.14	22.83	0.01	17.3 (18.06)
	5	30.14	33.34	0.001	21.7 (23.63)
	10	30.14	48.02	0.01	27.4 (29.40)
30	1	42.56	9.88	0.06	10.3 (9.88)
	2	42.56	18.14	0.02	15.6 (15.73)
	3	42.56	25.47	0.01	19.7 (20.40)
	5	42.56	38.16	0.001	25.6 (27.75)
	10	42.56	61.78	0.005	34.9 (39.29)

The area in between – moderate k – is of especial interest if one anticipates that only a small proportion of the populations have slipped. In this area, both the Bonferroni and multiple comparison fractiles are very conservative, and a clear need for a better approximation is apparent. This may be provided by the asymptotic extension of the tables of Appendix 3.

Finally, if one does not have prior information on the number of outliers, then a backward elimination stepwise procedure should be used. Only two relatively minor difficulties arise in adapting the theory of Chapter 5 to this problem. One is that the fractiles of $T_{n:i}$ discussed there and tabulated in Appendix 4 do not yet extend to the case of infinite ν. This is a defect which could be remedied by further computation, but in view of the observation in Chapter 5 that the α fractile of $T_{n-i+1:1}$ seems to be within $\alpha/4$ of the required fractile, one may quite reasonably simply use fractiles of $T_{n:1}$ as approximations. These fractiles are obtained to the required degree of approximation very simply from the $\alpha/(m-i+1)$ fractiles of the normal distribution.

The other difficulty is one that does not arise in the parametric case, and that is that every time one accepts that a treatment has not slipped and pools it with the inlying treatments, the ranks of the data in the inlying treatments alter. This situation does not arise in the parametric case where the values of the data cannot alter as a result of the sequence of hypothesis tests; however, in view of the fact that the Wilcoxon statistic is a function of the inlying treatments only through their grand mean rank, this alteration of the data probably does not introduce any noticeable difference in performance from the parametric case.

As mentioned earlier, the Mosteller test has also been adapted to the situation of multiple outliers. Neave (1975) suggests the following approach. Let the treatments be ranked in descending order of their largest order statistics. Let T_i be the number of observations in treatment 1 exceeding the largest order statistic in treatment 2. Then define $T_1 + T_2$ to be the total number of observations in treatments 1 and 2 exceeding the largest order statistic for treatment 3. Define in the same way $T_1 + T_2 + T_3$, $T_1 + T_2 + T_3, \ldots$, obtaining $T_2, T_3, T_4, \ldots$ by difference.

Neave gives the distribution of each T_i, and points out that they may be used to test for slippage of $k > 1$ populations. Of course, the actual testing procedure to be used is a by-now familiar problem of multiple decisions, and involves some additional difficulties apart from the distributional ones. The distribution is

$$\Pr[T_i > h_i | T_1, \ldots, T_{i-1}] = G(n_1 + n_2 + \ldots + n_i - t, n - t, h_i) \quad (6.3)$$

where $t = T_1 + \ldots + T_{i-1}$

and $G(a, b, h) = (a-1)!(b-h)!/(a-h)!(b-1)!$

Neave provides tables of G; however, it is a sufficiently simple function to be recomputed as and when required.

The statistics T_i are useful for testing whether exactly i populations have slipped. A rather different problem is that of deciding whether any slippage at all has occurred. One statistic for this is that of Granger and Neave (1968). They propose finding the top M order statistics of the total sample, and letting Y_i be the number of these from population i. Letting e_i be the expected value of Y_i under the null hypothesis ($e_i = Mn_i/n$) and a correction factor $A = (n-M)/(n-1)$, the test statistic is

$$S = \sum (Y_i - e_i)^2/(Ae_i) \quad (6.4)$$

which is approximately χ^2_{m-1}.

Neave (1973) presents some power simulations comparing a one-way Anova, the Kruskal–Wallis test, S, and the Mosteller test and some variants of it. The simulation was of normal data with one slipped population. The figures show:

(i) That in small samples, all methods are about equally effective;
(ii) That the ranking of the non-parametric tests (from best to worst) is Kruskal–Wallis, Granger-Neave, Mosteller.

As regards the latter conclusion, we may note that the Kruskal–Wallis test is especially effective for the logistic distribution (to which the normal distribution used in the simulation is a good approximation). Even given this caveat, however, it does seem that the earlier claim for the Doornbos over the Mosteller statistic is justified.

Two other statistics for multiple slippage are obtained by generalizing the Mosteller test. Bofinger (1965), for the problem of testing whether a specific k treatments have slipped, proposes the Mosteller statistic for the combined group of those k treatments against the rest of the samples. the paper contains the null distribution, an asymptotic form, and some power calculations.

Conover (1968) proposes a statistic somewhat like, but intuitively less attractive than the Neave T_i. This statistic involves first ranking the treatments as in the computation of the T_i. Then $M(i, j)$ is defined to be the number of observations in treatment i which exceed all observations in treatment j. A test for slippage of k treatments may be based on $M(k, k+1)$. We may note that the T_i and $M(i, j)$ are related by

$$T_1 + T_2 + \ldots + T_{i-1} = M(1, i) + M(2, i) + \ldots + M(i-1, i)$$

and so the T_i might be thought of as being functions of the $M(i, j)$:

$$T_i = \sum_1^i \{M(j, i+1) - M(j, i)\}$$

Furthermore, the Bofinger statistic for k groups is just $T_1 + T_2 + \ldots + T_k$, and so may be regarded as a single function of either the set of T_i or the set $M(i, j)$.

6.6 Large-sample outlier detection

A non-parametric test for the presence of outliers may be set up for a 'well-behaved' but otherwise unspecified distribution. Suppose one anticipates r outliers on the left. The statistic, due to Walsh (1958), rejects the null hypothesis if

$$X_{(r)} - (1 + A)X_{(r+1)} + AX_{(N)} < 0 \tag{6.5}$$

where N is the largest integer contained in $r + (2n)^{1/2}$. An obvious adaptation tests for r outliers on the right, and outliers on both sides may be tested using a combination of these two one-sided tests. We note, however, that the number of outliers anticipated on each side must be specified *a priori*.

The values of N and A are determined from the requirement that if $Z = X_{(r)} - (1 + A)X_{(r+1)} + AX_{(N)}$ then

$$E(Z) = B\{\text{var}(Z)\}^{1/2}\{1 + o(1)\} \tag{6.6}$$

for B prespecified. Then, in consequence of Chebyshev's inequality,

$$\Pr[Z < 0] = \Pr\left[\frac{Z - E(Z)}{\{\text{var } Z\}^{1/2}} < -B + o(1)\right] \leqslant \frac{1}{B^2} + o(1) \tag{6.7}$$

and so the size of the test does not exceed $1/B^2$. This then gives

$$A = \{1 + B[(c - B^2)(c - 1)]\}/(c - B^2 - 1) \tag{6.8}$$

where c is the largest integer contained in $(2n)^{1/2}$. The test is only applicable for samples sufficiently large that $c > B^2 + 1$.

For example, if $i = 2$, $B^2 = 10$, then a test of size 0.1 results from rejecting two smallest observations if

$$X_{(2)} - 3.348X_{(3)} + 2.348X_{(14)} < 0$$

The minimum sample size needed to set up a test at significance level 0.05 is $n = 221$. It may be seen from this that despite the very mild distributional requirements, the test has very severe sample-size requirements, which restrict its applicability.

A further difficulty is the assumption that r is specified *a priori*, though this assumption could, given enough data, be relaxed by the use of a Bonferroni inequality.

Another test (Walsh, 1950; 1953) which does not have such restrictive sample-size requirements can be set up if one knows *a priori* that the underlying distribution is symmetric. For this test too, the number of outliers must be specified *a priori*. If r outliers on the right are anticipated, one rejects H if

$$\min_{1 \leq k \leq s \leq r} [X_{(n+1-ik)} + X_{(jk)}] > 2X_{(W)} \qquad (6.9)$$

where $i_s = r, \quad i_u < i_{u+1}, \quad j_u < j_{u+1}, \quad j_s < W < n + 1 - r$

and W is the smallest integer satisfying

$$\Pr[X_{(W)} < \phi | \phi = \text{median}] \leqslant \alpha$$

Suitable values of the i, j and s are tabulated in Walsh (1953). An example with $r = 5, s = 2$, has $\alpha = 0.0547$. For this $j_1 = 1, j_2 = 2, i_1 = 4$, $i_2 = 5$. Letting U_α denote the α fractile of a standard normal distribution,

$$W = n/2 + U_{.0547}\sqrt{n/2}$$

and the test rejects H_0 in favour of a 5-outlier alternative if

$$\min [X_{(n-3)} + X_{(1)}, X_{(n-4)} + X_{(2)}] > 2X_{(W)}$$

It seems that there has been no detailed study of the sample sizes required for this test to have the specified asymptotic characteristics. However, the sample-size requirements appear not to be stringent.

Unlike the situation in non-parametric slippage testing, where the efficiency relative to the parametric test is known, nothing seems to be known about the relative performance of these two outlier tests. Both have however been shown to be consistent, with power tending to 1 as the outliers are moved further away from the main body of the data.

CHAPTER 7

Outliers from the linear model

7.1 The linear model

The standard form of the general linear model is

$$\mathbf{X} = \mathbf{W}\boldsymbol{\beta} + \boldsymbol{\varepsilon} \tag{7.1}$$

where the $n \times 1$ vector $\mathbf{X}$ represents the independent variables measured, the $n \times p$ matrix $\mathbf{W}$ is the design matrix, and $\boldsymbol{\beta}$ the $p \times 1$ vector of unknown regression coefficients. The error vector $\boldsymbol{\varepsilon}$ is assumed to consist of n independent, identically distributed $N(0, \sigma^2)$ variables.

Within the general linear model, it is sensible to distinguish two broad types of situation, distinguished by the design matrix $\mathbf{W}$. In the first, the regression situation, $\mathbf{W}$ consists of the values of several predictor variables which are measured along with $\mathbf{X}$, and which may be used to predict $\mathbf{X}$. As a rule, the elements of $\mathbf{W}$ take on many different values. The second situation is the analysis of variance situation in which the elements of $\mathbf{W}$ are either 0 or 1, and indicate membership in one of a set of disjoint classes. The formulation of such an analysis of variance may be illustrated by a very simple one-way analysis of variance, for which the model is a reformulation of

$$X_{ij} = \mu_i + \varepsilon_{ij} \qquad j = 1 \ \text{to} \ n_i, \qquad i = 1 \ \text{to} \ m$$

In both variants, outliers may be found – that is, individual X_{ij} whose true mean is not that indicated by the model, but some other quantity. The analysis-of-variance model, however, may also allow a 'slippage' formulation. In this, it is not isolated observations whose means are displaced, but entire groups: for example, for some i,

$$X_{ij} = \mu + \Delta + \varepsilon_{ij}$$

in which Δ represents the amount by which the ith population has slipped. In the case of a balanced analysis of variance (of whatever order) in which all cells have the same frequency, the slippage model presents no difficulty whatever. One merely condenses the observations down to the cell means and the within-cells sum of squared deviations, and proceeds to identify outliers amongst the means (see for example Chapter 3 and 5 and the material in the following sections). If the design is not balanced, however, the slippage problem becomes non-trivial, and in fact has not been solved in any satisfactory way. We will return to this problem later, concentrating attention for the moment on the outlier problem.

With data arising from a linear hypothesis, a difficulty arises which is not present in the more usual single-sample case. It is that, even if we know how many outliers are present, it is not easy, and may be impossible, to say which observations are erroneous. To illustrate this point, consider the case $p = 1$, and suppose that $n - 2$ of the data have $W_i = 0$ while the remaining two have $W_i = -1$ and $W_i = 1$ respectively. If an outlier occurs on either of the last two observations, then it is quite impossible to detect whether the error is in the last observation and is, say, of magnitude d, or is in the second last and of magnitude $-d$.

This case is admittedly extreme, but does bring out the severe difficulties that can arise when the outlier is associated with a design vector that deviates only in the directions that are associated with low eigenvalues of $\mathbf{W}'\mathbf{W}$. An entire class of problems has this ambiguity built in – that of analysis of variance in which one of the factors has only two levels. (It is advisable in this situation to eliminate the 50 per cent of redundant residuals immediately.)

Furthermore, when several outliers are present, similar difficulties arise with design matrices that are far from exotic. The origin and approach to solution of these problems will be set out in more detail later.

All information about outliers is contained in the vector of residuals

$$\begin{aligned}\mathbf{e} &= \mathbf{X} - \mathbf{W}\hat{\boldsymbol{\beta}} = \mathbf{X} - \mathbf{W}(\mathbf{W}'\mathbf{W})^{-1}\mathbf{W}\mathbf{X} \\ &= [\mathbf{I} - \mathbf{W}(\mathbf{W}'\mathbf{W})^{-1}\mathbf{W}']\mathbf{X} \\ &= [\mathbf{I} - \mathbf{W}(\mathbf{W}'/\mathbf{W})^{-1}\mathbf{W}']\boldsymbol{\varepsilon} = \mathbf{A}\boldsymbol{\varepsilon} \qquad (7.2)\end{aligned}$$

where

$$\mathbf{A} = \mathbf{I} - \mathbf{W}(\mathbf{W}'\mathbf{W})^{-1}\mathbf{W}'$$

and under the null hypothesis $\mathbf{e} \sim \mathrm{N}[\mathbf{0}, \sigma^2\mathbf{A}]$, a degenerate distribution, since the idempotent matrix $\mathbf{A}$ has rank $n - p$. One should note that even though all information about outliers is contained in $\mathbf{e}$, it does not follow that the individual elements of **e** are themselves much use in detecting outliers unless **A** is very well structured (for example, has all diagonal elements equal to one constant, and all off-diagonal elements to another).

The difficulties caused by an outlier are apparent: if any component of $\boldsymbol{\varepsilon}$ contains an aberration, this aberration will be 'smeared' by the matrix **A** over all elements of **e**.

No great difficulty of principle is involved in testing for a single outlier: the 'obvious' statistic is an analogue of the B^* statistic of Chapter 3, namely

$$B = \max |e_i| / \sqrt{(Sa_{ii})} \tag{7.3}$$

where $S = (\mathbf{X} - \mathbf{W}\hat{\boldsymbol{\beta}})'(\mathbf{X} - \mathbf{W}\hat{\boldsymbol{\beta}})$ and a_{ij} denotes the (i, j)th element of **A**.

The obvious statistic B does not have known optimality properties except in a rather restricted class of alternatives sketched in Hawkins (1976). The theoretical problems arise from the non-exchangeability of the elements of **e**.

The general distribution of $e_i/\sqrt{(Sa_{ii})}$ is given in several sources – notably Ellenberg (1973), where the joint distribution of k arbitrary scaled residuals $e_i/\sqrt{(Sa_{ii})}$ is given. The distribution of B will be discussed in Section 7.4

7.2 Recursive residuals and updating

Let us consider again the model $\mathbf{X} = \mathbf{W}\boldsymbol{\beta} + \boldsymbol{\varepsilon}$ written in partitioned form

$$\begin{bmatrix} \mathbf{X}_1 \\ X_n \end{bmatrix} = \begin{bmatrix} \mathbf{W}_1 \\ \mathbf{w}_n \end{bmatrix} \boldsymbol{\beta} + \begin{bmatrix} \boldsymbol{\varepsilon}_1 \\ \varepsilon_n \end{bmatrix} \tag{7.4}$$

where X_n, $\mathbf{w}_n$ is the last observation. Now if the nth observation were omitted from the sample, the summary statistics for the subregression would be

$$\hat{\boldsymbol{\beta}}_1 = (\mathbf{W}_1'\mathbf{W}_1)^{-1}\mathbf{W}_1'\mathbf{X}_1$$
$$S_1 = (\mathbf{X}_1 - \mathbf{W}_1\hat{\boldsymbol{\beta}}_1)'(\mathbf{X}_1 - \mathbf{W}_1\hat{\boldsymbol{\beta}}_1)$$

Now from Plackett (1950),

$$\hat{\boldsymbol{\beta}} = \hat{\boldsymbol{\beta}}_1 + (X_n - \mathbf{w}_n\hat{\boldsymbol{\beta}}_1)\mathbf{d}/(1 + c)$$
$$S = S_1 + (X_n - \mathbf{w}_n\hat{\boldsymbol{\beta}}_1)^2/(1 + c)$$
$$(X_n - \mathbf{w}_n\hat{\boldsymbol{\beta}}) = (X_n - \mathbf{w}_n\hat{\boldsymbol{\beta}}_1)/(1 + c)$$

where

$$\mathbf{d} = (\mathbf{W}_1'\mathbf{W}_1)^{-1}\mathbf{w}_n' \qquad (7.5)$$
$$c = \mathbf{w}_n\mathbf{d}$$

Furthermore, since

$$(\mathbf{W}'\mathbf{W})^{-1} = (\mathbf{W}_1'\mathbf{W}_1)^{-1} - \mathbf{dd}'/(1 + c)$$

these formulae may readily be adapted to the removal of X_n, $\mathbf{w}_n$ from the regression – that is, producing $(\mathbf{W}_1'\mathbf{W}_1)^{-1}$, $\boldsymbol{\hat\beta}_1$ and S_1 from $(\mathbf{W}'\mathbf{W})^{-1}$, $\boldsymbol{\hat\beta}$ and S.

Beckman and Trussell (1975) show that

$$r_n = (X_n - \mathbf{w}_n\boldsymbol{\hat\beta})/(1 + c)^{1/2} \qquad (7.6)$$

is distributed as $N(0, \sigma^2)$ and is independent of $\mathbf{X}_1$

Now clearly the operation of removing the observation X_n, $\mathbf{w}_n$ may be repeated on the matrices $\mathbf{X}_1$, $\mathbf{W}_1$ and another observation removed, r_{n-1} being formed *en passant*.

In this way, (assuming that $\mathbf{W}$ has full column rank p), one may produce $r_n, r_{n-1}, \ldots, r_{p+1}$, $n - p$ mutually independent $N(0, \sigma^2)$ variables. These are the so-called recursive residuals. They may also be defined in the following formal way: find an upper triangular matrix $\mathbf{B}$ having the property

$$\mathbf{BAB}' = \begin{pmatrix} \mathbf{0} & \mathbf{0} \\ \mathbf{0} & \mathbf{I} \end{pmatrix}$$

(The existence of such a matrix is a well-known result of matrix theory; to achieve the desired factorization may however require that some variables be interchanged: this problem is not especially germane to the immediate discussion.)

Defining the $n \times 1$ vector $\mathbf{R}$ by $\mathbf{R} = \mathbf{Be}$, we find that $r_1 = r_2 = \ldots = r_p = 0$; while $r_{p+1}, \ldots, r_n$ are the recursive residuals as defined before. This representation in terms of recursive residuals has important implications both for the theory and implementation of outlier tests in the general linear model.

The error sum of squares is $S = \Sigma_{p+1}^{n} r_i^2$. It is easily shown that if the last k observations (corresponding to the last k recursive residuals) are removed, then the residual sum of squares for the regression on this subvector is $S_k = \Sigma_{p+1}^{n-k} r_i^2$. Since we may have external information on σ^2 in the form of U, an independent $\sigma^2\chi_\nu^2$ variable, we will at once incorporate this information into both estimators, redefining

$$S = U + \sum_{p+1}^{n} r_i^2$$
$$S_k = U + \sum_{p+1}^{n-k} r_i^2 \tag{7.7}$$

The statistic $e_n/\sqrt{(Sa_{nn})}$ is easily shown to satisfy

$$e_n^2/Sa_{nn} = (S - S_1)/S \tag{7.8}$$

Suppose finally that in some way one can permute the observations so that the suspected outliers are in positions $n, n-1, \ldots, n-k+1$. Then obvious extensions of the statistics defined earlier for the normal distribution are $E_k = S_k/S$, a global statistic for the presence of up to k outliers; and $R_k = r_{n-k+1}/\sqrt{S_k}$, a stepwise statistic appropriate for testing for the presence of k outliers given that the sample contains $k-1$ outliers.

This glib formulation has side-stepped a major problem, which is specific to outliers from the linear model: this is that it is generally impossible to arrange the data in such a way that the suspected outliers are in order of aberration. This is because the set of suspect data given that k outliers are anticipated, need not overlap at all with that given that $k-1$ outliers are anticipated. Such a problem cannot arise in the case of independently and identically distributed data, except trivially when the number of outliers exceeds the number of good data.

For a simple example of a set of data for which the 'nesting' property does not hold, consider the set of data with $p = 1$:

$$\begin{array}{lll} W_i \approx 0, & X_i \approx 0 & i = 1 \text{ to } n-3 \\ W_{n-2} \approx -1 & X_{n-2} \approx 2 & \\ W_{n-1} \approx W_n \approx 1 & X_{n-1} \approx X_n \approx 2 & \end{array}$$

Told that this set of data contained one outlier, one would identify it as X_{n-2}, and then X_{n-1} and X_n would look entirely acceptable, both lying practically on the regression line. On the other hand, told that there were two outliers, one would identify them as X_{n-1} and X_n. Then X_{n-2} would look acceptable. The estimated slope of the regression line would of course be entirely different depending on which observations were rejected.

The unreliability of the nesting property means that stepwise procedures for outlier testing in the general linear model should be used with great caution. The statistic E_k is not affected by this possible non-operation of the nesting property; one must, however, use a comprehensive search to ensure that one has found the best set of outliers.

In an effort to overcome these difficulties and so to bring the problem down to the same level of complexity as is presented by a random sample, Bradu and Kass (1977) propose a novel solution. To decide whether a particular X_i, say X_1, is an outlier, take an arbitrary subset of size p of the $(X_i, \mathbf{w}_i)$ pairs, $i = 2, \ldots, n$. Use these to estimate $\boldsymbol{\beta}$, and find the recursive residual of X_1 corresponding to this estimate of $\boldsymbol{\beta}$.

By repeating this operation for all possible subsets of the $(X_i, \mathbf{w}_i)$ pairs, or for a large sample of them if n is so large as to make exhaustive enumeration unattractive, one can find the distribution of these recursive residuals for X_1. Now if there are a number of other outliers, but this number is small relative to n, then most of the subsets used will be 'clean', or nearly so. Thus comparing the distribution of recursive residuals for X_1 with those of $X_2, \ldots, X_n$ will show whether or not X_1 is an outlier. Specifically, such quantities as the quartiles of the distribution will be noticeably larger for contaminants than for inliers.

The obvious attraction of this method is that it is impossible for several outliers to mask each others' presence. While it has not yet been made the basis for a formal test, Bradu and Kass report that as a descriptive method it is very effective in identifying outliers, and with a firm indication of which observations are suspect, one is much better able to apply one of the formal tests for outliers set out below.

7.3 A regression formulation for outliers

The outlier model may itself be set up in terms of a regression. To see this, write the model as

$$\mathbf{X} = \mathbf{W}\boldsymbol{\beta} + \mathbf{I}\boldsymbol{\lambda} + \boldsymbol{\varepsilon} \tag{7.9}$$

where $\mathbf{I}$ is an $n \times n$ identity matrix and $\boldsymbol{\lambda}$ an $n \times 1$ vector of unknown parameters, most of whose elements we will require to be zero. Writing $\mathbf{V} = (\mathbf{W} \quad \mathbf{I})$

$$\gamma = \begin{pmatrix} \boldsymbol{\beta} \\ \boldsymbol{\lambda} \end{pmatrix} \tag{7.10}$$

the model is

$$\mathbf{X} = \mathbf{V}\gamma + \varepsilon \tag{7.11}$$

and is in the standard form for a general linear model, except for the rather unusual property that it seems to have $n+p$ unknown parameters and only n observations. The basic statistic necessary for fitting a regression is the matrix

$$(\mathbf{W} \quad \mathbf{X} \quad \mathbf{I})'(\mathbf{W} \quad \mathbf{X} \quad \mathbf{I}) = \begin{pmatrix} \mathbf{W}'\mathbf{W} & \mathbf{W}'\mathbf{X} & \mathbf{W}' \\ \mathbf{X}'\mathbf{W} & \mathbf{X}'\mathbf{X} & \mathbf{X}' \\ \mathbf{W} & \mathbf{X} & \mathbf{I} \end{pmatrix} \tag{7.12}$$

Using the conventional pivoting operations of stepwise regression to introduce the elements of $\boldsymbol{\beta}$ into the regression will transform this matrix to

$$\begin{pmatrix} -(\mathbf{W}'\mathbf{W})^{-1} & -(\mathbf{W}'\mathbf{W})^{-1}\mathbf{W}'\mathbf{X} & -(\mathbf{W}'\mathbf{W})^{-1}\mathbf{W}' \\ -\mathbf{X}'\mathbf{W}(\mathbf{W}'\mathbf{W})^{-1} & \mathbf{X}'\mathbf{A}\mathbf{X} & \mathbf{X}'\mathbf{A} \\ -\mathbf{W}(\mathbf{W}'\mathbf{W})^{-1} & \mathbf{A}\mathbf{X} & \mathbf{A} \end{pmatrix} \tag{7.13}$$

Concentrating on the lower right square submatrix of this matrix and using earlier identities, we see that it is simply

$$\begin{pmatrix} S & \mathbf{e}' \\ \mathbf{e} & \mathbf{A} \end{pmatrix} \tag{7.14}$$

Now the partial correlation which one would use to test whether, say, λ_n should be added to the regression, is obviously

$$r_n = e_n/\sqrt{(Sa_{nn})} \tag{7.15}$$

which is simply the Ellenberg statistic for testing X_n as an outlier. The fact that under the null hypothesis it is distributed like any other partial correlation provides an independent method of finding its marginal distribution, and shows that

$$\frac{(n+v-p-1)^{1/2}}{(1-r_n^2)^{1/2}} r_n \sim t_{n+v-p-1} \tag{7.61}$$

for an arbitrary scaled studentized residual r_n.

Next, if some λ – say λ_n – is introduced into the regression then we may update the matrix by sweeping on the row and column corresponding to λ_n. However, some mathematics confirms what common sense would dictate: if λ_n is introduced to the regression, then all the information contained in X_n will be absorbed in estimating λ_n and so the effect on $\hat{\beta}$, S and the remaining residuals must be identical to that of removing X_n, $\mathbf{w}_n$ from the data set. This fact provides a very interesting alternative way of looking at the computations of recursive residuals and updating.

The estimate of λ_n is $\hat{\lambda}_n = e_n/a_{nn} = X_n - \mathbf{w}_n\hat{\beta}_1$ which is interesting but obvious – it is the natural estimate of the amount of aberration of X_n if X_n is in fact an outlier.

As a matter of practical computation, the use of sweeping in the regression formulation to remove data from the sample would be wasteful – each sweep involves $O(n+p)^2$ operations, while the updating formulae, which provide the same information, involve only $O(p^2)$ operations. The regression formulation is therefore of interest only because of its theoretical consequences, and the insight it gives into some of the problems of multiple outliers. It shows that the problems associated with testing for multiple outliers are identical to those of identifying good subregressions, and suggests the possibility of adaptation of useful regression methods to the outlier problem. First, there is the relationship between the recursive residuals and the successive partial correlations in a regression problem. Even without the benefit of the earlier discussion of the masking effect, the analogy provides a warning of likely difficulties with forward selection, and this suggests that a backward elimination procedure may be better. The problem that the best set of k outliers may fail to include the best set of $k-1$ outliers is also familiar. Finally, the global assessment statistic $E_k = S_k/S$ is a monotonic transform of the squared partial multiple correlation coefficient used in regression to test whether a set of predictors contains any predictive power.

7.4 Distributional results

Consider the recursive residuals computed for an arbitrary ordering of the data. Then

$$E_k = S_k/S$$

and so

$$k(S - S_k)/\{S_k(v + n - p - k)\} = k(E_k^{-1} - 1)/(v + n - p - k) \quad (7.17)$$

follows an F-distribution with k and $v + n - p - k$ degrees of freedom. Now if in fact, E_k is computed for k outliers, then its value is the smallest of the $M = n!/k!(n-k)!$ statistics that could have been set up for testing k observations out of the n. Thus a conservative level-α test on E_k may be made using the upper α/M fractile of the indicated F-distribution. Of course, here, as in the earlier case, there is every reason to believe that for $k > 1$, these fractiles will be extremely conservative.

One interesting and potentially useful observation, however, is that the parameters ν and p seems to enter into the problem primarily as the term $\nu - p$. There is thus a good possibility that an adequate approximation to the fractiles of E_k as so defined may be obtained by using fractiles for the simple one-sample situation, using $\nu - p + 1$ rather than ν degrees of freedom for the external information on σ. This procedure is only applicable if ν exceeds $p - 1$.

A little heuristic reasoning soon convinces one that this approximation will lead to a conservative fractile for E_k. The reasoning is as follows: in the linear hypothesis, one might consider that $p - 1$ of the external degrees of freedom are 'traded off' against the $p - 1$ of the data points that are needed to estimate the additional $p - 1$ location parameters in the model. If one views the problem as one of succesive regression updating, it is clear that these $p - 1$ data points must always be selected from the inliers, and so the sum of squares that they account for must be stochastically smaller than that corresponding to the $p - 1$ external degrees of freedom. It thus follows that E_k is a stochastically increasing function of p, and that the fractiles from the simple random sample case will provide a conservative bound for those of the linear hypothesis.

How far out is this conservative bound? For the case $n = 20$, $p = 6$, $\nu = 5$, a simulation of 1000 random samples was carried out and the statistics E_2, E_3, E_4 and E_5 computed. The percentage of these smaller than the $\alpha = 0.10$ fractiles for $n = 20$, $p = 1$, $\nu = 0$ (the simple random sample case) was:

$k = 2$	percentage	9.6
3		8.7
4		7.9
5		5.9

Only for $k = 5$ might one regard the approximation as unduly conservative.

Even though this is a very limited simulation, it is clear what the general pattern is. If either n or ν increases, the approximation must improve since the source of conservatism – the difference between the two sums of squares on $p - 1$ degrees of freedom, becomes steadily less important. On the other hand, larger values of p will exacerbate the situation and increase the degree of conservatism.

For $k = 1$, the test based on recursive residuals reduces to one equivalent to the Ellenberg statistic B. The largest r_n is tested (largest $|r_n|$ for two-sided testing). Marginally,

$$t_i = (\nu + n - p - 1)^{1/2} r_i/(1 - r_i^2)^{1/2} \tag{7.18}$$

follows a t-distribution with $\nu + n - p - 1$ degrees of freedom. The largest of these then will exceed the α/n fractile of a t-distribution with $\nu + n - p - 1$ degrees of freedom with a probability not exceeding α.

In fact, as Srikantan (1961) shows, this approximation is exact if n, ν and α are sufficiently small. This fact represents an extension to the linear model of Pearson and Chandra Sekar's result and corresponds exactly to the use of the general approximation outlined above.

We have already noted that for some designs, the residuals are highly correlated, a result that has strong implications for the accuracy of this Bonferroni first bound. To take an extreme case, consider a 2^2 factorial design with σ known and equal to 1, say. Here, the four residuals are equal in absolute value, but the first Bonferroni bound behaves as if their common value is the largest of four such residuals. Thus the bound would be based on the $\alpha/4$ fractile of a N(0, 1) distribution instead of the α fractile, and so the approximation would be terrible.

Before exploring the question of more accurate approximations, let us first note that a lower bound can be obtained quite easily. Any resolution into recursive residuals $r_n, \ldots, r_{p+1}$ yields $n - p$ independent N(0, σ^2) variables, in which by construction r_n cannot exceed the largest $e_i/\sqrt{a_{ii}}$ in absolute value. Thus the statistic B is stochastically larger than max $|r_i|/\sqrt{S}$, whose α fractile may be obtained to an excellent degree of approximation from the $\alpha/(n-p)$ fractile of a beta distribution with degrees of freedom $\frac{1}{2}$ and $\frac{1}{2}(\nu + n - p - 1)$. This means that the true α fractile of B is bracketed by the upper $\alpha/(n-p)$ and the α/n fractiles of the same beta distribution. Any outlier statistic exceeding the larger of these fractiles is certainly significant at the α level of significance, and any below the smaller is certainly not significant, and so only in the area between the two is there any doubt about about significance. Thus if the two bounds are close together (as is the case whenever n is large relative to p) then there is little point in pursuing the question of better approximations to the fractiles.

Better approximations, and even exact values, may in principle be obtained by the use of further terms in the Boole expansion, but these further terms, unlike the first, depend on the matrix A. The necessary joint distribution theory of the e_i is set out in Ellenberg (1973), and its application to the evaluation of the successive terms of the Boole expansion may be found in Stefansky (1971, 1972).

It is clear that to get fractiles corresponding to every possible matrix A would be totally impracticable. However, if all diagonal elements of A are equal to one constant, and all off-diagonal elements to another, then the computations become more feasible. This condition is met by factorial experiments and balanced analyses of variance generally. Stefansky provided bounds to the true fractiles using the first, and the first two terms of the Boole expansion for a number of two-way layouts. For layouts of between 3 and 9 rows and columns, she shows that these upper and lower bounds are either identical, or disagree by only a small amount. The practical implication of this, that the first Bonferroni bound provides a very good approximation for the two-way layout, is extended to a variety of other factorial designs by John and Prescott (1975).

Thus it appears that the first Bonferroni bound should provide good approximations in all but the most extreme cases of interdependence amongst residuals. Tables of fractiles based on this approximation may be found in Appendix 10.

For the case $\nu = 0$, $p = 2$, Tietjen, Moore and Beckman (1973) observe that the α fractile of the largest $|t|$ is well approximated by the upper $\alpha/2$ fractile of $(X_{(n)} - \bar{X})/\sqrt{S}$ based on a sample of a size $n - 1$. In the light of the above comments, we recognize this observation as saying that the $\alpha/2(n+1)$ fractile of a given t-distribution is well approximated by the $\alpha/2n$ fractile: certainly not a strong or surprising statement.

The stepwise distribution of the successive studentized recursive residuals $r_i = e_i/(\Sigma_{p+1}^{i-1} e_j^2)^{1/2}$ when the recursive residuals are expected in some sort of order of aberration is not known. As with E_k one might hope quite reasonably that it would approximate that of the simple case with $\nu - p + 1$ external degrees of freedom. In fact $E_k = \prod_1^k (1 + r_i^2)^{-1}$, so the fact that the random-sample case yields approximate fractiles which become steadily more conservative as k increases implies that a similar relationship should hold between the $T_{n:i}$ of Chapter 5 and the statistics $r_i/(1 + r_i^2)^{1/2}$ used in the linear hypothesis.

7.5 Identification of multiple outliers

Let us now return to the problem mentioned earlier – that of actually identifying the k outliers – an apparent prerequisite for testing whether in fact they deviate significantly from the remaining $n-k$. This problem reduces to that of finding the best subset of k predictors out on n given the multiple regression defined by the matrix

$$\begin{pmatrix} S & \mathbf{e}' \\ \mathbf{e} & \mathbf{A} \end{pmatrix}$$

Since the $n \times n$ matrix $\mathbf{A}$ has rank only $n-p$, it goes without saying that the matrix $\mathbf{W}$ is, in the usual regression sense, ill-conditioned, and so difficulties about alternative equally good subsets of predictors are quite likely.

The first approach is the generation of the optimal subset of size k for each k using a branch-and-bound algorithm. Furnival and Wilson (1974) present a highly efficient method of doing this which allows the full set of $2^p - 1$ partitions to be scanned very efficiently, and in an amount of computer time which would be acceptable even for n of the order of 30. In fact, given the equivalence of adding one term to a regression and updating, there seem to be good prospects for further and more specific adaptations of general branch-and-bound theory.

The next possibility is the use of some stepwise method together with a check on the likelihood of there being alternative subsets. One such approach is the following. Set an upper limit K on the number of outliers to be detected, and introduce K of the λ_i using forward selection. Then remove them all, using backward elimination, taking out the least significant λ_i at each stage. The sequence of 't' statistics in the removal stage gives the quantities necessary for formal testing for the presence and number of outliers.

Two checks are made along the way to see whether any other set of the X_i would do as well as a set of outliers. The first check is the changes in partial correlations between X and each λ_i not included in the regression. If the introduction or removal of any λ_i causes an abrupt change in the partial correlation for some other λ_i, then there is likely to be some indeterminacy as to which of the observations are outliers.

The sceond check applies to the various observations included as outliers at each stage. Following Scott (1975), we note that for each λ_i included in the regression, a Student t may be computed. Let t_m^2 denote the mean of these individual t^2 values. The overall significance of the regression may also be measured by means of the F ratio, and, for orthogonal predictors, $F = t_m^2$. Scott's conditioning statistic is $c = (t_m^2/F - 1)/(t_m^2/F + 1)$. Its value is near zero in well-conditioned problems, and Scott suggests that a value of 0.33 divides well-conditioned from ill-conditioned problems.

This statistic tells whether, amongst those λ_i include in the regression, there are problems of alternative subsets. The check on the changes in partial correlations similarly shows up substitutabilities between the λ_i that are included and those that are excluded. Thus a sensible procedure seems to bc to accept the results of the backward elimination if neither check suggests the likelihood of alternative subsets. If either does suggest alternative subsets, then it is wise to use an all-subset, or a branch-and-bound approach to indicate all possible subsets of aberrant observations.

7.6 Example

Andrews (1974) discusses a set of data that are most interesting from the point of view of outlier detection. The data, which are reproduced in Table 7.1, relate to the conversion of ammonia to nitric acid.

TABLE 7.1

X	W_1	W_2	W_3
42	80	27	89
37	80	27	88
37	75	25	90
28	62	24	87
18	62	22	87
18	62	23	87
19	62	24	93
20	62	24	93
15	58	23	87
14	58	18	80
14	58	18	89
13	58	17	88
11	58	18	82
12	58	19	93
8	50	18	89
7	50	18	86
8	50	19	72
8	50	19	79
9	50	20	80
15	56	20	82
15	70	20	91

X = Stack loss
W_1 = Air flow
W_2 = Cooling water inlet temperature
W_3 = Acid concentration

A conventional probability plot of residuals suggests that observation 21 may be an outlier, but gives no reason to suspect any other observation, or that any assumption for the linear model is not satisfied.

The data were re-analysed using the theory described earlier in this chapter. For ease of operation, a stepwise regression formulation was used rather than a deletion routine so that the predictors W_1, W_2 and W_3 could be introduced and eliminated at will. The predictor W_3 was found to be uniformly non-significant, and so was omitted from all regressions. Rather interestingly, W_2, initially and subsequently significant, became non-significant once two outliers had been eliminated; nevertheless, it was retained in all regressions to avoid muddying the water unnecessarily.

The results of recursively eliminating five outliers are shown in Table 7.2. The choice of specifically five outliers was arbitrary. The test statistics are shown as the familiar Student's t-statistics used in regression rather than as $T_{n:i}$ to facilitate their intuitive interpretation. The format of the table needs some explanation: the outliers tested were, in order, observations 21, 4, 3, 1 and 13. The entries on and above the diagonal of Table 7.2 show the t value for each of them as further outliers were deleted. We note that these t values increase as additional outliers are removed and the error variance is correspondingly reduced.

The entries below the diagonal give the t value for each observation before the outlier for that stage was removed. For example, after removal of X_{21} but just before removing X_4, X_{21} had a t value for reinclusion into the sample of -4.64, while X_4, X_3, X_1, and X_{13} had t values for elimination from the sample of 3.36, 1.48, 1.11 and -1.04 respectively. The values below the diagonal are uniformly non-significant. We have already mentioned that a probability plot of the residuals only suggests the presence a single outlier; the low t values at

TABLE 7.2 t values for outlier rejection

Observation	Stage 1	2	3	4	5
21	−3.47	−4.64	−4.76	−5.64	−6.73
4	(2.03)	3.36	3.89	6.00	6.90
3	(1.65)	(1.48)	2.26	4.37	4.78
1	(1.38)	(1.11)	(1.86)	4.05	4.47
13	(0.83)	(−1.04)	(−1.57)	(−1.62)	−2.25

the subsequent stages show that even as one deletes outliers, the residuals at any stage never suggest the presence of more than one further outlier.

The design matrix is not good. The Scott conditioning statistic for the regression of X on W_1 and W_2 is 0.63, and over the five stages of outlier removal it becomes successively 0.60, 0.56, 0.61, 0.67 and 0.66; all well in excess of Scott's recommended maximum of 0.33. There may thus be other summaries of the data than the one presented here in terms of regression on W_1, W_2 and the outliers indicated.

Finally, consider the inferential aspects of the data. Note that at stage 3, the t value is only 2.26 and well below significance at any reasonable level. Even the initial t of -3.47 corresponds to a size of 6.8 per cent, so that a strict approach would lead to one's not rejecting any outliers at all by forward selection.

To assess the backward elimination test depends on fractiles not available. However, since there are only three predictors used, one may reasonably use the fractiles for $T_{n:i}$ with $n = 20$, $\nu = 0$ as an approximation not seriously in error. The sequence of fractiles for 5 outliers and $\alpha = 0.05$ is 3.60, 3.67, 3.65, 3.64 and 3.56. We thus conclude at the 0.05 significance level that observations 1, 3, 4 and 21 are outliers, while the remainder are inliers; a conclusion in agreement with that reached by Andrews using robust regression followed by a subjective decision as to whether values were inliers or outliers.

7.7 The general slippage problem

Let us now return to the slippage problem. The general problem is illustrated quite well by a one-way analysis of variance

$$Y_{ij} = \mu_i + \varepsilon_{ij} \qquad j = 1 \text{ to } n_i, \qquad i = 1 \text{ to } m$$

The simplest (and rather typical) slippage problem is

$$\begin{aligned} \mu_i &= \mu + \Delta && \text{for some unknown } i \\ \mu_k &= \mu && \text{for all } k \neq i \end{aligned}$$

We may reduce the original data to the sufficient statistics

$$X_i = \sum_j Y_{ij}/n_i \qquad i = 1 \text{ to } m$$

$$U = \sum_i \sum_j (Y_{ij} - X_i)^2$$

with no loss of information, thereby setting the problem within the general framework of Chapter 3.

In the case that all n_i are equal, the solution to the slippage problem is well known, and uses the statistic $(X_{(m)} - X)/\sqrt{S}$. However, with unequal n_i, no such optimality results are known. The regression approach of seeking the λ_i most highly correlated with X is clearly equivalent to one seeking to minimize the weighted residual variance of the X_i retained in the sample. This is turn in equivalent to using as test statistic $\max\sqrt{n_i}|(X_i - \bar{X})|/\sqrt{S}$ where $\bar{X} = \Sigma n_i X_i/\Sigma n_i$. This test statistic is a monotonic function of the 'two-sample t statistic' for comparing the set Y_{ij}, $j = 1$ to n_i, with the remaining $n - n_i$ observations.

All of these aliases describe a quite reasonable procedure, though the conditions under which it is optimal are not known.

7.8 Slippage test performance

The obvious application of the slippage test is to a one-way analysis of variance, in which the alternative hypothesis specifies that one of the means is slipped relative to the others. This might be a reasonable model for such situations as a comparative study of several treatments in which it is anticipated that one treatment should prove markedly different from the others.

One method of analysis is a standard one-way analysis of variance. Now as shown by Scheffe (1958, p. 68) the conventional Anova F-test is equivalent to the following multiple comparison. Given $X_1, \ldots, X_m$ believed independently and identically distributed as $N(\xi, \sigma^2)$ and independent of $U \sim \sigma^2\chi^2_\nu$ (to which the balanced one-way Anova may be reduced), consider all contrasts $\Sigma a_i X_i$ with $\Sigma a_i = 0$. Then, simultaneously over all possible contrasts, with probability $1 - \alpha$,

$$|\sum a_i\{X_i - E(X_i)\}| < \hat{\sigma}b, \ \hat{\sigma}^2 = U/\nu$$

where

$$b^2 = (\sum a_i^2)(m-1)F_{m-1,\nu,1-\alpha}$$

the latter denoting the upper α fractile of an F-distribution with $m - 1$ and ν degrees of freedom.

Furthermore, the null hypothesis of all $E(X_i)$ being equal is rejected by the Anova if and only if at least one contrast is significant.

Now suppose that the correct alternative is in fact one of slippage with the first population slipped by an amount Δ and we will assume that Δ is large enough to ensure that if an outlier test is significant, it will correctly identify X_1 as the outlier with very high probability, rather than some other X_j.

The contrast of particular interest is $(X_1 - \bar{X}_1)$ and we are interested in the probability under H_1 that $D > D_F$ where $D = (X_1 - \bar{X}_1) \times \{v(m-1)/(mU)\}^{1/2}$ and D_F is its multiple comparison fractile

$$D_F^2 = (m-1)F_{m,v,1-\alpha}$$

Now one alternative test is the use of the Nair (1948) procedure, whose test statistic is $(X_{(m)} - \bar{X})/\sqrt{U}$. Under the alternative hypothesis with Δ large enough to ensure that there is negligible probability of mis-identifying an outlier, the probability of a correct decision by this procedure is easily seen to be $\Pr[D > D_N]$ where, by Bonferroni's inequality, D_N is well approximated by

$$D_N^2 = F_{1,v,1-\alpha/m}.$$

Finally, one may use the slippage test. This has a probability of correct decision (from Chapter 3) of $\Pr[T > T_S]$ where

$$T = (X_1 - \bar{X}_1)\left\{\frac{(m-1)(m+v-2)}{m(U+V)}\right\}^{1/2}$$

where $V = S_1$ is a $\sigma^2\chi^2$ variable with $n-2$ degrees of freedom and is clearly independent of $\bar{X}_1$ and X_1. Using earlier results again, T_S is well approximated by

$$T_S^2 = F_{1,m+v-2,1-\alpha/m}$$

How do the performances of these criteria compare? We note first that since the Scheffe procedure allows more general contrasts than are used in the Nair procedure, it must be true that $D_F > D_N$, and so the multiple comparison procedure must perform worse than Nair's statistic.

Next, since the Pearson–Chandra Sekar statistic is an optimal slippage statistic, it must be more powerful than the Nair statistic.

Under the alternative hypothesis, it is easy to show that D and T follow noncentral t-distributions with degrees of freedom v and $m+v-2$, and the same noncentrality $\delta = \Delta\{(m-1)/m\}^{1/2}$.

The relative superiority of T is obviously greatest when v is small. To illustrate this, and the comparative performance, some approximate figures follow. These were calculated using the Wilson–Hilferty approximation to the F-distribution and the Johnson–Welch approximation to the noncentral t-distributions.

The first figures are for $m = 10$ and $v = 10$. This corresponds to a balanced Anova with 10 classes and 2 replicates per class. The approximate fractiles and powers are listed in Table 7.3, demonstrat-

TABLE 7.3 $m = 10, \nu = 10$

	Fractile	Power $\Delta = 4$	Power $\Delta = 5$
Anova	5.2	0.21	0.44
Nair	4.0	0.49	0.77
Optimal	3.5	0.66	0.90

ing a clear superiority for the two slippage tests, and the optimal one in particular.

Leaving $m = 10$ but increasing ν to 40 yields the results in Table 7.4. Here, the superiority of the optimal test to the Nair statistic is much smaller, but both are much superior to the multiple comparison procedure.

This discussion has concentrated on a slippage test for a single slipped population. If the number of slipped populations is not known *a priori* then a less impressive increase in power using slippage tests is possible. Suppose, for example, that one anticipates at most three populations which are assumed, for convenience, to have slipped by the same amount. Since the amount of slippage is assumed to be the same for the slipped populations, take the largest of the two-sample t statistics made up of no more than three observations in one group. Using the very conservative Bonferroni approximation, we note that this t is the largest of $10 + 45 + 120 = 175$ possible t values. Using the 0.05/175 fractile of a t-distribution then reduces to fractiles of 4.6 for $n = 10$, $\nu = 10$ and 4.0 for $n = 10$, $\nu = 40$. Both these values are still considerably smaller than would follow from multiple comparisons. Hence, even though both are highly conservative, they would yield greater power.

As a further possibility, suppose that the number of slipped populations is left entirely unspecified, but it is specified that all have the same mean. Then one may proceed as in automatic interaction

TABLE 7.4 $m = 10, \nu = 40$

	Fractile	Power $\Delta = 4$	Power $\Delta = 5$
Anova	4.4	0.37	0.71
Nair	3.2	0.76	0.95
Optimal	3.2	0.77	0.96

detection by finding the largest t value for all possible partitions of the k groups into two. Asymptotically for $m = 10$ $\nu \to \infty$, this leads to a 0.05 fractile for the largest t of 3.5 (see, for example, Kass, 1975). Even this value is smaller than the value of 4.11 that would be used in a multiple comparison test.

The use of other outlier statistics such as the sequential tests or those based on criteria such as the E_k offers exciting possibilities for capitalizing on the partial information that even under the alternative hypothesis it may still be reasonable to assume that a majority of the true means are equal.

As this comment implies, there is little or no comparative work in the literature on the relative performance of Anova and more general slippage tests. Further study should pay particular attention to the effect of a mis-specification of the actual number of slipped populations, since the outlier tests can lose power as a result of masking.

7.9 Exploring interactions

In a multiway analysis of variance, the slippage test may be put to an entirely different use. This is the use described in Brown (1975) though his approach is somewhat different from that used here. Brown observes that the discovery of a significant interaction in a multiway analysis of variance can obstruct interpretation considerably, since it indicates merely that there is some violation of additivity somewhere in the Anova table. One possible model of this violation might well be that the table is additive except for some small number of individual cells whose true means have slipped from the values implied by additivity. By setting up a slippage test for such cells, one can bring the analysis to a very satisfactory conclusion by identifying the location and amount of non-additivity.

This approach may also be used in analysing fractional designs which do not normally provide a test of any interactions. For example, the 15 degrees of freedom of a Latin Square for 3 factors at 4 levels yield 6 degrees of freedom for 'error' – actually an aggregate of second- and third-order interactions, together with pure error. An analysis to see whether any of the 16 cells has slipped from additivity can provide a very useful check on these interactions – one that is all but impossible to make on the basis of a study of the raw residuals, which are very highly intercorrelated.

CHAPTER 8

Multivariate outlier detection

The concepts of parametric outlier testing extend with some difficulty but only relatively minor modification to multivariate data. Suppose that $\mathbf{X}_1, \mathbf{X}_2, \ldots, \mathbf{X}_n$ are n vectors of p components, the null hypothesis being that they are a random sample from the multivariate normal distribution with mean vector ξ and covariance matrix Σ

$$H_0 : \mathbf{X}_i \sim \mathrm{MN}(\xi, \Sigma) \qquad i = 1 \text{ to } n \tag{8.1}$$

Under the alternative hypothesis of, say, k outliers, there is some unknown permutation $j(i)$ of the integers $1, 2, \ldots, n$ such that

$$\mathbf{X}_{j(i)} \sim \mathrm{MN}(\xi, \Sigma) \qquad i = k+1 \text{ to } n \tag{8.2}$$

while $\mathbf{X}_{j(1)}, \ldots, \mathbf{X}_{j(k)}$ follow some other distribution or distributions. It is convenient, and seems not to result in much loss of generality, to suppose that the outliers also follow multivariate normal distributions, but with at least one parameter different from (ξ, Σ).

The major possibilities that come to mind are

$$\begin{aligned}
&H_1 : \mathbf{X}_{j(i)} \sim \mathrm{MN}(\xi_i, \Sigma) \qquad i = 1 \text{ to } k\\
&H_{1a} : \xi_1 = \xi_2 = \ldots = \xi_k\\
&H_2 : \mathbf{X}_{j(i)} \sim \mathrm{MN}(\xi, a_i\Sigma) \qquad i = 1 \text{ to } k\\
&H_{2a} : a_1 = a_2 = \ldots = a_k\\
&H_3 : \mathbf{X}_{j(i)} \sim \mathrm{MN}(\xi, \Sigma_i) \qquad i = 1 \text{ to } k,\\
&H_{3a} : \Sigma_1 = \Sigma_2 = \ldots = \Sigma_k
\end{aligned} \tag{8.3}$$

In each case, the specialization involved in assuming that the outliers come from the same distribution reduces the problem essentially from a $(k+1)$-sample to a two-sample problem.

Some care is needed to select the alternatives in such a way that they are sensible models for outliers. Outliers are values with high probabilities of occurring where the probability density of the true distribution is low, remote from the main body of data. Viewed in this light, we see that for any $\xi_i \neq \xi$, model H_1 will describe outliers whose probability density is higher under H_1 than under H_0 in some 'tail' regions.

Under H_0, the probability density of an observation is

$$f(\mathbf{x}) = K \exp - (Q/2)$$

where $$Q = (\mathbf{x} - \xi)'\Sigma^{-1}(\mathbf{x} - \xi), \quad K = \{(2\pi)^p |\Sigma|\}^{1/2} \tag{8.4}$$

The density is thus small where Q is large, and vice versa. Under H_2, the ith contaminant has density

$$f(\mathbf{x}) = K a_i^{p/2} \exp - Q/(2a_i)$$

So if $a_i > 1$, then the density is higher under H_2 than under H_0 wherever Q is sufficiently large – that is, outliers are likely to occur in any 'tail' region.

Conversely, $a_i < 1$ would specify a contaminating distribution giving rise to inliers, since the density is more concentrated near $\mathbf{x} = \xi$. Thus only the situation $a_i > 1$ is of interest for specifying outliers.

The more general model H_3 gives rise to some extra difficulties, and seems not to have been considered in the literature. Letting

$$Q_0 = (\mathbf{x} - \xi)'\Sigma^{-1}(\mathbf{x} - \xi)$$
$$Q_1 = (\mathbf{x} - \xi)'\Sigma_i^{-1}(\mathbf{x} - \xi)$$

and $T = \Sigma^{-1} - \Sigma_i^{-1}$, standard results in matrix algebra show that

$$T \text{ positive definite} \Rightarrow Q_0 - Q > 0$$
$$T \text{ negative definite} \Rightarrow Q_0 - Q < 0$$

while if T is indefinite, $Q_0 - Q_1$ is positive along some rays and negative along others.

The first situation, T positive definite, will lead to greater densities in all the tail regions under H_3 then under H_0. Thus model H_3 could be used to describe a process of contamination leading to outliers. If T is negative definite, then the density under H_3 is more concentrated near ξ than under H_0, and so contaminants will be inliers. The final situation, T indefinite, corresponds to Σ and Σ_i having different shapes. The alternative density tends to yield outliers in certain directions (those directions in which $Q_0 - Q_1 > 0$) and inliers in others (those in which $Q_0 - Q_1 < 0$).

Probably because of the difficulties this leads to, the model H_3 is not popular, and attention has focused on the models H_1 and H_2.

8.1 General testing principles

It is a general problem in multivariate, as opposed to univariate statistics, that most situations with composite alternative hypotheses do not lead to unique optimal test statistics unless additional constraints on the alternative hypothesis, or the class of statistics studied, are imposed. Because of this, univariate statistics can often be generalized in more than one way, each of the generalizations being optimal under a particular set of conditions.

As in the general situation, outlier testing on multivariate samples is complicated by the fact that univariate tests generalize in more than one way, each generalization having some optimality properties.

Consider first the case of at most a single contaminant. The sufficient statistics under H_0 are

$$\bar{\mathbf{X}} = \sum_i \mathbf{X}_i/n$$

and

$$\mathbf{A} = \sum_i (\mathbf{X}_i - \bar{\mathbf{X}})(\mathbf{X}_i - \bar{\mathbf{X}})' \tag{8.5}$$

or

$$\mathbf{S} = \mathbf{A}/(n-1)$$

If $\mathbf{X}_i$ is known to be a contaminant then clearly sufficient statistics are

$$\bar{\mathbf{X}}_i = \sum_{i \neq j} \mathbf{X}_j/(n-1)$$

and

$$\mathbf{A}_i = \sum_{i \neq j} (\mathbf{X}_j - \bar{\mathbf{X}}_i)(\mathbf{X}_j - \bar{\mathbf{X}}_i)' \tag{8.6}$$

or

$$\mathbf{S}_i = \mathbf{A}_i/(n-2)$$

Note that external information, in the form of $\mathbf{U} \sim W(\boldsymbol{\Sigma}, v)$, an independent Wishart-distributed variable, can be accommodated quite easily – the statistics $\mathbf{A}$ and $\mathbf{A}_i$ are replaced with $\mathbf{A} + \mathbf{U}$ and $\mathbf{A}_i + \mathbf{U}$.

Conventional theory of the two-sample tests (see, for example, Press, 1972) suggests the following single-outlier statistics:

(i) Two-sample T^2 test for H_0 against H_1

$$T_i^2 = (\mathbf{X}_i - \bar{\mathbf{X}}_i)'\mathbf{A}_i^{-1}(\mathbf{X}_i - \bar{\mathbf{X}}_i)(n-1)(n+v-2)/n$$

(ii) Wilks's lambda statistic for the two-group multivariate analysis of variance test for H_0 against H_1.

$$\Lambda = |\mathbf{A}_i|/|\mathbf{A}|$$

Using the algebraic identity

$$\mathbf{A} = \mathbf{A}_i + (n-1)/n(\mathbf{X}_i - \bar{\mathbf{X}}_i)(\mathbf{X}_i - \bar{\mathbf{X}}_i)'$$

and standard results on matrices, we soon see that

$$\begin{aligned}\Lambda^{-1} &= 1 + (n-1)(\mathbf{X}_i - \bar{\mathbf{X}}_i)'\mathbf{A}_i^{-1}(\mathbf{X}_i - \bar{\mathbf{X}}_i)/n \\ &= 1 + T_i^2/(n+\nu-2)\end{aligned}$$

and so (as is well known) the Wilks's lambda test is exactly equivalent to the T^2 test.

While specific optimality properties for the statistic $T^2_{\max} = \max_i T_i^2$ are not at once apparent, it is well known that the two-sample test T^2 is uniformly most powerful unbiased for testing H_0 against H_1 in the class of tests that are invariant under arbitrary full-rank linear transformation. From this it seems that $T^2_{\max}$ will, as in the case with its univariate equivalent, maximize the probability of a correct decision, in the single outlier problem, amongst the class of unbiased tests invariant under full-rank linear transformations.

Of course, it is legitimate to ask whether it is sensible to restrict oneself to tests invariant under all full-rank linear transformations. In some cases, it is not, and we will see how, in these cases, a test whose performance is better than that of $T^2_{\max}$ may exist.

The model H_2 might be derived from H_1 in the following way, which is reminiscent of either a Bayesian or a random-effects approach

$$\begin{aligned}&\mathbf{X}_{j(i)}|\xi_i \sim \mathrm{MN}(\xi_i, \Sigma) \\ \text{while} \quad &\xi_i \sim \mathrm{MN}(\xi, \{a_i - 1\}\Sigma)\end{aligned} \tag{8.7}$$

which leads to the marginal density

$$\mathbf{X}_{j(i)} \sim \mathrm{MN}(\xi, a_i\Sigma).$$

Thus from the optimality properties under H_1, one might deduce that $T^2_{\max}$ is also a good statistic for testing H_0 against H_2.

In fact, Ferguson (1961a) showed that, in the class of tests invariant under full-rank linear transformations, the test maximizing the probability of correct decision uses as statistic

$$D^2_{\max} = \max_i D_i^2 \qquad \text{where } D_i^2 = (\mathbf{X}_i - \bar{\mathbf{X}})'\mathbf{A}^{-1}(\mathbf{X}_i - \bar{\mathbf{X}})$$

However, some quite straightforward matrix manipulations (e.g. Press, 1972, p. 23) show that

$$D_i^2 = T_i^2/(n+\nu-2+T_i^2) \tag{8.8}$$

and so the statistics D_i^2 and T_i^2 are equivalent – a result with exact parallels in the univariate case. (The mathematical equivalence should not obscure the fact that computationally D_i^2 is much preferable to T_i^2, since the same coefficient matrix, $\mathbf{A}^{-1}$, is used in all quadratic forms.) The identity connecting D_i^2 and T_i^2 is familiar in the context of discriminant analysis, where it is used in the opposite direction for jack-knifing.

Thus we find that, if the restriction to tests invariant under arbitrary full-rank linear transformation is acceptable, then for both H_1 and H_2 the optimal outlier test uses as statistic $T^2_{\max}$.

The satisfactorily tidy solution for the single outlier situation does not extend to the multiple outlier case. Suppose first that the model H_1 is believed to apply. If one knew the permutation $j(i)$, and this permutation were, say, $j(i) = i$, then the problem would be that of testing H_0 against the alternative

$$\mathbf{X}_i \sim \begin{cases} \mathrm{MN}(\xi_i, \Sigma) & i = 1 \text{ to } k \\ \mathrm{MN}(\xi, \Sigma) & i = k+1 \text{ to } n \end{cases}$$

In the absence of any assumption about the identity of some or all of the ξ_i, this is simply a $k+1$ group one-way multivariate analysis of variance (Manova) set-up. Sufficient statistics are

$$\mathbf{X}_i \qquad i = 1 \text{ to } k$$

$$\bar{\mathbf{X}}_k \text{ (say)} = \sum_{j=k+1}^{n} \mathbf{X}_j/(n-k)$$

$$\mathbf{A}_{(k)}, \text{ (say)} = \sum_{j=k+1}^{n} (\mathbf{X}_j - \bar{\mathbf{X}}_k)(\mathbf{X}_j - \bar{\mathbf{X}}_k)'.$$

while under H_0 the sufficient statistics are $\bar{\mathbf{X}}$ and $\mathbf{A}$.

In the univariate case, the test statistic applied would be an F-test. For the Manova equivalent, however, there are several possible tests. To define these various tests and show their interconnections more clearly, suppose that the ith eigenvalue of $\mathbf{A}\mathbf{A}_{(k)}^{-1}$ is $1 + \lambda_i$, $i = 1, \ldots, p$. Then common criteria (Press, 1972, p. 253) include

(i) Wilks's lambda,

$$\Lambda = \prod_1^p (1 + \lambda_i)^{-1}$$

This is the generalized likelihood ratio test, and apart from the good power against all alternatives common to members of its class, has the advantage of good, well-known approximations to its null distribution.

(ii) Roy's largest-root criterion λ, the largest of the λ_i. This criterion is optimal if the vectors $\xi, \xi_1, \ldots, \xi_k$ are collinear, or nearly so. Its fractiles are usually read from Heck charts (Heck, 1960), but these are not suitable for outlier testing, which requires fractiles of unusual significance levels. A partial solution to this problem may be found in the use of an algorithm for evaluating the distribution function of λ (Venables, 1974). Unfortunately, this algorithm is not completely general: $n-k-p$ must be odd, though even values may be treated approximately by interpolation.

(iii) Hotellings $T_0^2 = \Sigma_1^p \lambda_i$, which also has the alternative expression

$$T_0^2 = \sum_1^k (\mathbf{X}_i - \bar{\mathbf{X}})' \mathbf{A}_{(k)}^{-1} (\mathbf{X}_i - \bar{\mathbf{X}}) + (n-k)(\bar{\mathbf{X}}_k - \bar{\mathbf{X}}) \mathbf{A}_{(k)}^{-1} (\bar{\mathbf{X}}_k - \bar{\mathbf{X}})$$

General exact distributional results on T_0^2 are not available, which limits its usefulness. Furthermore, despite its appealing form as the sum of several quadratic forms, there is no evidence of any common circumstances in which it is noticeably more powerful than Λ (Ito, 1962), and so its use is not recommended in general.

The null hypothesis is rejected for small values of Λ, and large values of λ and T_0^2. If it is not known *a priori* which of the n observations are outliers, then one computes the test statistic for all $M = n!/k!(n-k)!$ partitions of the data into k outliers and $n-k$ inliers. The most extreme of these M values is used as the outlier statistic, and the corresponding partition is used to identify the outliers.

Let us denote by Λ_k the value of Λ for a k-outlier model. Wilks (1963) proposed the use of Λ_k as a k-outlier test statistic and noted that, but for the ranking of the $\mathbf{X}_i$, Λ_1 and Λ_2 have quite straightforward distributions. Specifically, Λ_1 would follow a beta distribution with parameters $(n-p-1)/2$ and $p/2$, while Λ_2 follows a beta distribution with parameters $n-p-2$ and p.

More general distributional results are set out in Anderson(1959) where it is shown that Λ_k may be expressed in terms of the product of $(k+1)/2$ beta variables. Alternatively, the Box approximation (Press, 1972) may be used for $k > 2$.

Returning to the outlier problem, Wilks suggested that Λ_k be evaluated and tested for significance at the α/M level of the null distribution. This suggestion provides the usual conservative test based on Bonferroni's inequality. As the corresponding results from the univariate case indicate, the bound for $k = 2$ is probably very conservative but is certainly better than no fractile at all.

Using the Bonferroni bound, Wilks provides tables of approximate fractiles for $p = 1(1)5$ and a range of n values from 5 to 500, and $k = 1,2$. Outside the range of these tables, one might use the exact F distributions for $k = 1$, 2 or $p = 1$, 2, or the Box approximation. Wilks's tables of these Bonferroni approximate fractiles are listed in Appendix 8.

Siotani (1959) considers a multivariate equivalent of the Nair statistic, namely $(\mathbf{X}_i - \mathbf{X})'\mathbf{U}^{-1}(\mathbf{X}_i - \bar{\mathbf{X}})$ where $\mathbf{U}$ is independent of the sample. By finding the second term in the Boole expansion, Siotani shows that the approximation provided by the Bonferroni inequality is an excellent one. It is reasonable to suppose that the same will apply to the Wilks outlier statistic when $k = 1$.

Another interesting but unexplored area is the possibility of using the largest-root criterion as a test statistic. As already noted, this is preferable to Λ_k if the true means $\xi, \xi_1, \ldots, \xi_k$ are approximately collinear. One can envisage quite easily situations in which this may apply. The variational interpretation of λ as corresponding to the smallest possible univariate E_k statistic attainable by transforming the data vectors $\mathbf{X}$ to scalars Y by premultiplying by a $1 \times p$ vector $\mathbf{c}$ gives it considerable appeal. One could get conservative Bonferroni approximations to the fractiles of this statistic in much the same way as using the Wilks approach.

8.2 Alternative approaches

Provided one is willing to restrict attention to statistics which are invariant under arbitrary full-rank linear transformations of the data, one's attention focuses naturally on functions of the eigenvalues $\lambda_1, \ldots, \lambda_p$. If this restriction is not made, however, then other interesting possibilities come to mind.

Amongst the more interesting of these is the use of suitable functions of the principal component residuals. To define these suppose that

$$\mathbf{X}_i \sim \mathrm{MN}(\xi, \Sigma)$$

in which it is usually sensible to scale so that Σ is in correlation form.

Let the $p \times p$ matrix $\mathbf{C}$ reduce Σ to canonical form

$$\mathbf{C}\Sigma\mathbf{C}' = \Gamma = \text{diag}(\gamma_i) \tag{8.9}$$

where $\mathbf{CC}' = \mathbf{C}'\mathbf{C} = \mathbf{I}$ and $0 < \gamma_1 < \ldots < \gamma_p$ are the eigenvalues of Σ ranked, we should note well, from smallest to largest.

Defining $\mathbf{Y}_i = \mathbf{C}(\mathbf{X}_i - \xi)$, $\mathbf{Y}_i$ is the vector of principal component residuals of $\mathbf{X}_i$. From the usual multivariate theory $\mathbf{Y}_i \sim \text{MN}(0, \Gamma)$, and so has independent elements.

Gnanadesikan and Kettenring (1972) point out that these principal component residuals have appealing properties for detection of certain types of outliers. Suppose for example that two components of $\mathbf{X}$ are highly correlated, but an observation yields values on these components that, while individually plausible, are not consonant with this high correlation (see the point marked 'o' in Fig. 8.1a). Such an outlier is described by Gnanadesikan and Kettenring as obscuring the near multicollinearity, and it will be identified by having a large principal component residual on one of the first few components of $\mathbf{Y}$. Figure 8.1b by contrast shows an outlier that inflates variances and covariances, and which will show up as a large principal component residual on one of the last few components of $\mathbf{Y}$. Thus the principal component residuals are interpretable, since, depending on which residual is large, one can identify the type of outlier present. Furthermore, since the principal component residuals are scalars, and are distributed independently under the normal model, one may apply univariate outlier tests to them.

Gnanadesikan and Kettenring do not propose any formal test to be applied to the components of $\mathbf{Y}$ to detect outliers. They recommend graphical exploratory methods such as bivariate plotting of the components of $\mathbf{Y}$. They also suggest gamma plots of the sums of

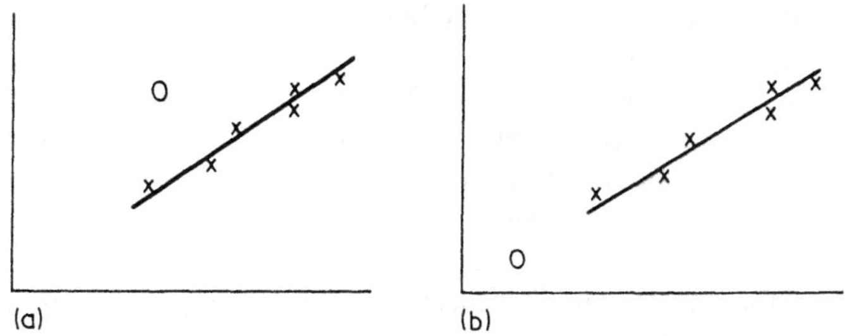

Figure 8.1

squares of the principal component residuals $\mathbf{Y}_i'\,\mathbf{Y}_i$ as a graphical method of locating outliers. This plot, unfortunately, is not a good way of locating outliers like that in Fig. 8.1a, for reasons which will appear below.

Hawkins (1974) suggests a rescaling of $\mathbf{Y}$ to Z defined by

$$\mathbf{Z}_i = \mathbf{\Gamma}^{-1/2}\mathbf{Y}_i$$

$$= \mathbf{B}(\mathbf{X}_i - \xi)$$

where

$$\mathbf{B} = \mathbf{\Gamma}^{-1/2}\mathbf{C}$$

or

$$b_{ij} = c_{ij}/\sqrt{\gamma_i}$$

From the well-known variational properties of the reduction to canonical form one can deduce the following property of **Z**:

Amongst all scalar linear transforms $\mathbf{cX}$ of $\mathbf{X}$ having unit variance, the choice $c_j = b_{1j}$ leads to a maximum value for $\Sigma_j c_j^2$. Continuing in this way, amongst all scalar linear transforms $\mathbf{cX}$ of $\mathbf{X}$ which are uncorrelated with $\Sigma_j b_{ij} X_j$ for $i = 1$ to $I - 1$ and have unit variance, the choice $c_j = b_{Ij}$ leads to a maximum value for $\Sigma_j c_j^2$.

The practical significance of this variational property may be seen if we suppose that some vector I is an outlier because it contains an error in a single unknown component J, i.e.,

$$\mathbf{X}_i^* \sim \mathrm{N}(\xi, \Sigma) \qquad i = 1 \text{ to } n$$

$$X_{ij} = X_{ij}^* \qquad i \neq I, \text{ or } j \neq J$$

$$X_{IJ} = X_{IJ}^* + \delta$$

Suppose one decides to seek this outlier by setting up a scalar normal variate $\mathbf{c}(\mathbf{X} - \xi)$ which has variance 1. Clearly this quantity will be maximally sensitive to the outlier if the elements of $\mathbf{c}$ are large numerically and so the property of the components of Z of maximizing the sum of squares of $\mathbf{c}$ makes them especially attractive for detecting outliers of this type.

Possible test statistics based on $\mathbf{Z}_i$ include max $|Z_{ij}|$, $(\Sigma_j Z_{ij}^2)$, and to locate the index i corresponding to the outlier, one would use the maximum of these statistics over i. The range of j in each case may be $1, 2, \ldots, p$ (in which case the latter statistic is identical to $(\mathbf{X}_i - \xi)'\,\Sigma^{-1}(\mathbf{X}_i - \xi)$); or $1, 2, \ldots, K$, $K < p$.

The Gnanadesikan–Kettenring statistic $\mathbf{Y}'\mathbf{Y}$ is equivalent to $\Sigma_j \gamma_j Z_{ij}^2$. Since the Z_{ij} corresponding to small γ_j are given a greatly reduced weight in this sum, we see (as was claimed earlier) that the statistic is not sensitive to outliers that obscure a near multicollinearity.

That a choice of $K < p$ is desirable is quite clear from theoretical considerations, for exactly analogous to the optimal property of the first component of **Z**, we have the property:

Amongst all scalar linear transforms **cX** of **X** having unit variance, the choice $c_j = b_{pj}$ leads to a minimum value for **cc**′′

Thus, in the same sense as before, the scaled principal component residuals corresponding to the large γ_i are the worst possible linear functions of **X** for detecting outliers on a single component, and so little is lost by discarding them. Hawkins (1974) offers some suggestions for solving the resulting problem of deciding how many components to retain.

8.3 Distribution theory

Since the components of **Z** are independent N(0, 1) variates, the distribution of the two outlier statistics $T_1 = \max_{i,j}|Z_{ij}|$ and $T_2 = \max_i(\Sigma_j Z_{ij}^2)$ are quite straightforward. If there are n data vectors and K components of **Z** are retained, then T_1 is the largest in absolute value of nK independent N(0, 1) variates, and so

$$\Pr[T_1 < t] = [\Phi(t)]^{nK}$$

where $\Phi(.)$ denotes the standard normal distribution function.

Also $\Sigma_1^K Z_{ij}^2$ is a χ_K^2 variate. Thus T_2 is the largest of n independent χ_K^2 variates, and so its fractiles may be found in the same fashion as

$$\Pr[T_2 < t] = (\Pr[\chi_K^2 < t])^n$$

In the noncentral case in which outliers are present, the Z_{ij} are still normal with unit variance, but not all have zero mean. By a direct computation of the mean, the performance of T_1 and T_2 may be deduced quite easily. Some actual numerical computations may be found in Hawkins (1974).

In order to make the transformation to **Z**, one must know ξ and Σ. In actual practice, if these parameters are not known, one must estimate ξ by $\bar{\mathbf{X}}$ and Σ by **S**, the sample statistics. Corresponding to the result $\mathbf{Z} \sim N(\mathbf{0}, \mathbf{I})$, one then finds that the $\mathbf{Z}_i$ have sample mean zero and sample covariance matrix **I**, but unknown distribution. While the theory for ξ and Σ known will evidently hold asymptotically as $n \to \infty$, there is evidence in Hawkins (1974) that unless p is very small, the values of n required for invocation of this asymptotic theory may be prohibitively large. Most of the difficulties result from estimation of Σ rather than of ξ. Thus it may be that, despite the appealing properties of the tests based on principal component residuals, they are not generally useful for formal testing with controlled probability of type I error, as their null hypothesis distributions are not known for samples of small to moderate size. The exception to this broad statement is the Wilks statistic $T_2 = \max_i(\Sigma_1^p Z_{ij}^2)$ whose distribution is known. Proceeding to the Wilks statistic via the Z_{ij} aids one's understanding in that if one finds a significantly large value of T_2, one can then inspect the individual Z_{ij} to establish what sort of outlier has occurred.

One should also not minimize the value of the Z_{ij} as descriptive indicators of the presence and nature of outliers in the sample, even if they are used only informally and not as the basis of some formal test of controlled size.

Despite this comment, however, the exact distribution of the principal component residuals is a matter of some interest in both outlier and other contexts, and some exact distributional results would be of great interest.

CHAPTER 9

Bayesian approach to outliers

We have already mentioned the suggestion by Glaisher that the usual arithmetic mean was not appropriate for observations having a common mean, but different precisions. Writing the model

$$X \sim \mathrm{N}(\xi, \sigma_i^2)$$

we note that an alternative to Glaisher's suggestion of an iterative approach to the estimation of the weight w_i to attach to X_i would be to specify a Bayesian model for the different σ_i. One approach might specify that the σ_i^2 are independent random variables from a common distribution. By choosing the inverted gamma for this distribution, we arrive at a model in which the $X_i - \xi$ form a random sample from a Student's t-distribution. This fact is of some interest in interpreting this commonly used, heavy-tailed family of distributions.

An outlier model, on the other hand, could be set up by specifying that the majority of the σ_i were equal while a few were larger. A fuller development of the Bayesian approach has been slow in appearing. This fact is especially surprising in view of the widespread use of Dixon's mechanism (ii) for outliers, in which each observation may be a good observation with probability p and an outlier with probability $1 - p$; a model that begs for posterior probabilities of outliers.

Much of the concern in the papers to date on the Bayesian approach has been focused on the problem of estimation when outliers may be present, while a relatively minor amount has been devoted to the question of whether, *a posteriori*, the sample actually does contain significant outliers.

De Finetti (1961) presented a Bayesian model of the type just sketched. Assuming that $\sigma_i^2 = \sigma^2 \gamma_i$, where the γ_i are independently and identically distributed with density $\alpha(.)$, say, de Finetti shows that the Bayes estimator of ξ is $\Sigma \rho_i X / \Sigma \rho_i$, where the ρ_i are given by intractable n-fold multiple integrals involving α and the X_i. Unfortunately, de Finetti does not work through the details of any particular special cases, nor does he address the question of how to test whether in fact all γ_i are equal to (i.e. whether any outliers are, *a posteriori*, evident in the data).

One formulation for dealing with the problem of testing for the presence of outliers is as follows. Let the basic distribution be $f_0(x|\theta)$ and the contaminating distribution $f_1(x|\theta, \delta)$. If k contaminants are present but it is not known which observations they are, then the likelihood of the sample may be written

$$L(\mathbf{X}|\theta, \boldsymbol{\delta}, k) = \left\{\prod_{i=1}^{n} f_0(X_i|\theta)\right\} \Sigma_I \left\{\prod_{i=1}^{k} f_1(X_{j(i)}|\theta, \delta_i)/f_0(X_{j(i)}|\theta)\right\} \tag{9.1}$$

where $j(1), \ldots, j(k)$ is some permutation of k from the integers $1, \ldots, n$, and the summation Σ_I runs over all possible permutations of k indices from n.

A special distinct case arises if the parameters δ_i in the likelihood are assumed equal; we will assume different δ_i. The posterior distribution of $\boldsymbol{\delta}$ (which is our primary concern) is then given by

$$g(\boldsymbol{\delta}|\mathbf{X}, k) \propto \int L(\mathbf{X}|\theta, \boldsymbol{\delta}, k) \cdot g(\theta, \boldsymbol{\delta})\, d\theta \tag{9.2}$$

If $f_0(x|\theta)$ is the special case of $f_1(x|\theta, \delta)$ when $\delta = \delta_0$, say, then a study of $g(\boldsymbol{\delta}|\mathbf{X}, k)$ near $\boldsymbol{\delta} = \boldsymbol{\delta}_0$ will provide evidence as to whether or not there is evidence of k contaminants.

Particular cases of this general model which have appeared in the literature are:

(i) $f_1(x|\theta_1, \theta_2, \delta) = \mathrm{N}(\theta_1 + \delta, \theta_2)$
with $f_0(x|\theta_1, \theta_2) = f_1(x|\theta_1, \theta_2, \delta = 0)$

This is the basis of Guttman's model, which will be considered below.

(ii) $f_1(x|\theta, \delta) = (\delta\theta)^{\alpha} x^{\alpha - 1} \exp(-\delta\theta x)/\Gamma(\alpha)$
with $f_0(x|\theta) = f_1(x|\theta, \delta = 1)$

This is Lingappaiah's model for the gamma distribution: α is assumed known.

The computational effort involved in evaluating L or any derived quantity (such as the posterior distribution) is prohibitive if k is at all large, because of the need to evaluate all the partitions making up the sum Σ_I.

9.1 Example

Take the model

$$f_1(x|\theta, \delta) = \delta\theta \exp(-\theta\delta x) \qquad 0 < \delta < 1$$

with $f_0(x|\theta) = f_1(x|\theta, \delta = 1)$

and suppose one has a sample $X_1, X_2, \ldots, X_n$. Then it follows on integrating out θ that the posterior densities for δ are:
One outlier

$$g(\delta|\text{data}) \propto \sum_j \delta/\{S - (1-\delta)X_{(j)}\}^{n+1}$$

where $S = 1 + \sum_1^n X_i$
Two outliers

$$g(\delta_1, \delta_2) \propto \sum_k \sum_l \delta_1\delta_2/\{S - (1-\delta_1)X_{(k)} - (1-\delta_2)X_{(l)}\}^{n+1}$$

in which a uniform prior over (0, 1) is assumed for δ.

To illustrate these densities for good and bad data, two artificial 'samples' were generated. One, sample A, consisted of the expected order statistics for a sample of size 10 from the standard exponential distribution. The other, sample B, was made up of the expected order satistics of a sample of size 8 from the standard exponential distribution, together with the two contaminating values 6 and 7, which should indicate the presence of two outliers. Under the single outlier model, the two samples yielded the following posterior densities for δ:

	$g(\delta)$	
δ	Sample A	Sample B
0.1	0.50	0.75
0.2	0.79	1.04
0.3	0.96	1.12
0.4	1.06	1.12
0.5	1.12	1.09
0.6	1.14	1.05
0.7	1.16	1.02
0.8	1.17	1.00
0.9	1.17	0.99
1.0	1.18	0.98

In sample A, the maximum density occurs at $\delta = 1$, which is as it should be, since the sample contains perfect data. The density is, however, rather flat for $\delta > 0.5$, showing that one might have a certain amount of unease about declaring no outliers. Sample B indicates outliers rather unequivocally and suggests a value of about 0.35 for δ. This value is not at all implausible; the value 7 is 1/0.4 times the expected maximum order statistic in a sample of size 10.

The bivariate densities for the two outlier model are (apart from the normalisation constant) as follows:

Sample A

	δ_1				
δ_2	0.2	0.4	0.6	0.8	1
0.2	0.44	0.57	0.60	0.61	0.61
0.4	0.57	0.74	0.80	0.81	0.81
0.6	0.60	0.80	0.86	0.87	0.90
0.8	0.61	0.81	0.87	0.89	0.90
1	0.61	0.81	0.88	0.90	0.90

Sample B

	δ_1				
δ_2	0.2	0.4	0.6	0.8	1
0.2	1.48	1.29	1.09	0.99	0.96
0.4	1.29	1.26	1.13	1.06	1.03
0.6	1.09	1.13	1.05	0.99	0.97
0.8	0.99	1.06	0.99	0.94	0.92
1	0.96	1.03	0.97	0.92	0.90

Here the data for sample A are fairly clear about the absence of outliers, while those of sample B indicate a strong pair of outliers – the maximum value of $g(\delta_1, \delta_2|\text{data})$ occurs near (0.2, 0.2). Some other features are interesting. For example, going down the $\delta_1 = 1$ column, we see that δ_2 is best estimated by a value near 0.4 rather than 0.2. The explanation for this is quite straightforward. If $\delta_1 = 1$ then the data contain only one outlier and then the other large value is regarded as an indication of the value of θ. This implies a lower value of θ and correspondingly a larger value for δ than under the double-outlier model. A similar but less marked tendency is also seen for the $\delta_1 = 0.6$ and 0.8 columns.

9.2 A 'test' for outliers

Another approach using the posterior distribution of $\boldsymbol{\delta}$ is the use of the Dixon contamination model, according to which, *a priori*,

$\Pr(k \text{ outliers}) = \binom{n}{k} p^k(1-p)^{n-k}$, so that, *a posteriori*,

$$\Pr(k \text{ outliers}|\text{data}) = \int \Pr(\text{data}|k, \theta, \boldsymbol{\delta}) \cdot g(\theta, \boldsymbol{\delta}) \, d\theta \, d\boldsymbol{\delta}. \Pr(k \text{ outliers}) \tag{9.3}$$

all of whose terms are now specified. To take a simple illustrative example with fully specified parameters, suppose that $f_0(x)$ is N(0, 1) while $f_1(x)$ is N(1, 1). Let $n = 5$, $p = 0.1$ and consider the 'sample' -1, 0, 1, 2, 3.

$$L(\mathbf{X}|k \text{ outliers}) \propto \sum_I \prod_1^k \exp\left[-\tfrac{1}{2}\{(X_{j(i)} - 1)^2 - X_{j(i)}^2\}\right]$$

$$= \sum_I \exp\left(\sum_i X_{j(i)} - k\right)$$

The 'sample' then yields

k	$\Pr(k)$	$\Sigma_I \exp(\Sigma_i X_{j(i)} - k)$	$\Pr(k\|\text{data})$
0	0.590	1	0.08
1	0.328	11.61	0.52
2	0.073	35.83	0.36
3	0.008	35.83	0.04
4	0.000	11.61	0.00

That is, there is clear posterior evidence of either 1 or 2 outliers. This conclusion is not especially sensitive to the value of p specified; only if p drops below 0.016 will the data favour 0 outliers rather than 1.

Despite appearances, this simple problem involved some fairly tedious calculations, and more realistic ones involving non-trivial distributions with unknown parameters and larger samples are considerably more laborious. For this reason, emphasis in the Bayesian area, like that in the classical one, has concentrated on the case of at most one outlier in the sample. Even this apparently easy problem involves quite noticeable problems of distributions and inference.

Gebhardt (1964) studies the following specialization of the de Finetti model: all observations X_i are independent, with $X_i \sim N(\xi_i, \sigma_i^2)$. Under the null hypothesis, all ξ_i are equal to ξ, while the σ_i all equal 1. (Note the restriction of the model to known σ). Two possible outlier models are that for some unknown i, $H_1 : \sigma_i > 1$ or $H_2 : \xi_i = \xi + a$ where a is assumed known. Complicated expressions for the Bayes estimator of ξ emerge.

Guttman (1973) proposed a model which was more realistic than the Gebhardt one, but which turns out both theoretically and computationally to be quite tractable. Assume that $n - 1$ of the X_i are $N(\xi, \sigma^2)$ while the remaining one is $N(\xi + a, \sigma^2)$. Using the common diffuse priors on the parameters ξ, a and σ, one then finds the posterior distribution of ξ, σ and a. In particular, that of a, the measure of deviation of the contaminant, is

$$f(a|\mathbf{X}) = \sum_1^n c_i h(a|\eta_i, B^{(i)}, n-2)$$

where

$$\eta_i = n(X_i - \bar{X})/(n-1)$$
$$B^{(i)} = [nA^{(i)}/(n-1)(n-2)]^{-1}$$
$$A^{(i)} = \sum_{j \neq i} (X_j - \bar{X}_i)^2$$
$$\bar{X}_i = \sum_{j \neq i} X_j/(n-1)$$
$$c_j = (A^{(j)})^{-(n-2)/2} \Big/ \sum_1^n (A^{(i)})^{-(n-2)/2}$$

and

$$h(w|\eta, B, v) = \frac{B^{1/2}\Gamma\{\frac{1}{2}(v+1)\}}{(\pi v)^{1/2}\Gamma\{\frac{1}{2}v\}}\left\{1 + \frac{B(w-\eta)^2}{v}\right\}^{-(v+1)/2} \tag{9.4}$$

That is, $f(a|\mathbf{X})$ is a finite mixture of scaled Student's t-densities, each with $n-2$ degrees of freedom. The jth component of this mixture centres at η_j, which we can recognize intuitively as an estimator of a if it is known that X_j is the contaminant. The weight c_j of the component is a decreasing function of $A^{(j)}$, and is easily interpreted: if X_j deviates markedly from the other data, then $A^{(j)}$ will be much smaller than the other $A^{(i)}$, and hence c_j will be large.

This reasoning at once identifies the c_j as providing a measure of how outlying X_j is. Of course, since c_j is proportional to the $(n-2)$ power of Grubb's classical single outlier statistic, it should be seen as an old friend in new guise, and not as a new outlier statistic.

The formal test for whether an outlier is present may be approached from the posterior density of a. If no outlier is present, one would expect to find

$$\Pr[a < 0|\mathbf{X}] = \Pr[a > 0|\mathbf{X}]$$

The formal odds ratio $\Pr[a > 0|\mathbf{X}]/\Pr[a < 0|\mathbf{X}]$ may thus be used as a 'test statistic' for the presence of an outlier, any values markedly different from 1 leading to the conclusion that an outlier is present.

Letting $G(t) = \int_{-\infty}^{t} h(x|0, 1, n-2)\mathrm{d}x$ denote the cumulative distribution function of the t density with $n-2$ degrees of freedom, this odds ratio is easily seen to be

$$\sum_{j=1}^{n} c_j G(n_j\sqrt{B^{(j)}}) \Big/ \sum_{j=1}^{n} c_j\{1 - G(\eta_j\sqrt{B^{(j)}})\} \qquad (9.5)$$

Guttman suggests that an odds ratio greater than 5 or less than 0.2 be taken to indicate that an outlier is present.

9.3 Example

A well-studied set of data is Herndon's measurements of the vertical semi-diameter of Venus. The sample values, and the corresponding c_j (in parentheses) are:

– 0.30(.0097)	– 0.44(.0117)	1.01(.0522)
0.48(.0118)	– 0.24(.0092)	0.06(.0083)
0.63(.0157)	– 0.13(.0085)	– 1.40(.8193)
– 0.22(.0090)	– 0.05(.0083)	0.20(.0087)
0.18(.0086)	0.39(.0104)	0.10(.0083)

The values – 1.40, and, to a lesser degree, 1.01 are suspected to be outliers. Certainly the value – 1.40 with its c_j of 0.8193 seems suspicious. The odds ratio is $0.121/0.879 = 1/7.26$, and, following Guttman's suggestion, we would conclude that an outlier is present. We may do more than that: since $\Pr[a < 0|\mathbf{X}] = 0.879 > 0.5$, we may conclude that an outlier is on the left, and quite confidently (in view of $c_j = 0.8193$) identify it as the value – 1.40.

The brief presentation of these results should not mislead the reader into believing that the calculations involved are negligible: they are in fact quite tedious.

The B^* statistic is 0.6704, and is not quite significant at the 5 per cent level. This is just an indication that the odds-ratio test is less stringent than the classical test and (in the classical sense) will have a higher type I error rate.

It is not at once clear from Guttman's paper how well this procedure is likely to fare in the presence of multiple outliers. The clue to this lies in the fact that $A^{(j)}$ is proportional to the Grubbs statistic, and so this Bayesian model, like the classical ones, will be affected by masking. The effect of the masking is to increase all $A^{(i)}$, so that no $c_{(i)}$ is much larger than the other c's, and to reduce the values of $\eta_j\sqrt{B^{(j)}}$ so that the posterior odds ratio is not significant.

Thus for cases where more than one outlier is suspected it is imperative that the algebraic obstacles in the path of the multiple outlier model be overcome.

It is of interest to note also that the Guttman model does not make use of the Dixon approach, namely that there are no outliers with probability $(1 - p)^n$, and outliers with the complementary probability.

The univariate model also extends in a quite natural way to the situation in which the $\mathbf{X}_i$ are $p \times 1$ vectors. Here the place of the Student's t-distributions is taken by a multivariate t, and the c_j now become functions of generalized variances. The formal odds-ratio test on the vector $\mathbf{a}$ is no longer possible, but tests equivalent to the univariate test may be made componentwise. The exact details of this parallel those of the univariate case quite closely, and may be found in Guttman (1973).

A somewhat different model which allows for multiple outliers, but contains no test, is due to Box and Tiao (1968). The starting point for this model is the linear hypothesis

$$\mathbf{X} = \mathbf{W}\boldsymbol{\beta} + \varepsilon$$

where with probability

$$p^{(k)} \propto p^k(1 - p)^{n-k}$$

any particular subset of k of the ε_i are from $N(0, b\sigma^2)$, the remaining ε_i being $N(0, \sigma^2)$ $(b > 1)$. From this model, it is possible to estimate $\boldsymbol{\beta}$. The parameter b is assumed known, but it is shown that the estimator is not very sensitive to the value specified for either b or p – a very fortunate discovery.

While the procedure does not involve a formal test such as the odds-ratio test in the Guttman procedure, it does contain an expression for the posterior probability that any particular observations are outliers. This expression could be used to provide a formal basis for testing for the presence of several outliers.

A Bayesian estimation model for the generalized gamma distribution

$$f(x) = bx^{\alpha-1}\beta^{\alpha/b}\exp(-\beta x^b)/\Gamma(\alpha/b)$$

has been proposed by Lingappaiah (1976). While this model is capable of handling multiple outliers, it contains no statistic which indicates how many outliers are actually present. However, since it is just a particular case of the general theory set out earlier, a test would be devised in much the same way as was illustrated for the normal case.

CHAPTER 10

Miscellaneous topics

10.1 Discrete distributions

The general procedure for testing for outliers in a continuous distribution with unknown parameters is: find a suitable (e.g. complete sufficient) statistic T for the parameters, and then find a suitable outlier test statistic $h(X, T)$ whose distribution does not depend on the unknown parameters.

This procedure does not work with discrete distributions, since the necessary pivotal quantities $h(X, T)$ do not exist. The standard procedure used is to condition the data on the value of the sufficient statistic. The conditional distribution (which involves no nuisance parameters) may then be used directly to test for outliers. To illustrate the general procedure, suppose the basic distribution is a member of the generalized power series distribution (GPSD) family.

$$f(x|\theta) = a(x)\theta^x / A(\theta)$$

This family includes the Poisson, binomial and negative binomial families, and many others less familiar.

Given observations $X_1, \ldots, X_n$, the sufficient statistic for θ is $T = \Sigma X_i$. The probability generating function of the X_i is $A(\theta t)/A(\theta)$, and that of T is $\{A(\theta t)/A(\theta)\}^n$ (Johnson and Kotz, 1969). From this, one may easily deduce the distribution of T, and that of the set $\{X_i\}$ given $T = t$, say.

Let $h(x|t)$ denote the conditional distribution of a single arbitrary X_i given $T = t$. Then a conservative outlier test using Bonferroni's inequality will reject $X_{(n)}$ if

$$\sum_{x = X_{(n)}}^{\infty} h(x|t) \leqslant \alpha/n \qquad (10.1)$$

This procedure is exactly analogous to those derived for the various continuous distributions studied.

Alternatively, one may evaluate the exact distribution of $X_{(n)}$ given

$T = t$. This result is analogous to the recursive expressions given earlier for the normal and gamma distributions. Following Hawkins (1976) analogously to Equation (2.9), let

$$f_n(x|t) = \Pr[X_{(n)} = x | T = t]$$

$$F_n(x|t) = \sum^x f_n(y|t)$$

Then

$$f_n(x|t) = \sum_{i=1}^{n} (-1)^i \binom{n}{i} F_{n-i}(x, t - ix) \prod_{j=0}^{i-1} h(x|t - jx, n - j) \quad (10.2)$$

where $h(x|t, n)$ denotes $h(x|t)$ when the sample is of size n. For example:

(i) Poisson distribution. Here,

$$h(x|t) = t!(n-1)^{t-x}/n^t x!(t-x)! \quad (10.3)$$

that is, the X_i are conditionally binomial $b(t, n^{-1})$.

(ii) Binomial distribution with $X_i \sim b(r, \theta)$. Here the required conditional distribution is hypergeometric.

$$h(x|t) = \binom{r}{x}\binom{nr - r}{t - x} \bigg/ \binom{nr}{t} \quad (10.4)$$

(iii) Negative binomial distribution

$$f(x|\theta) = \binom{-r}{x}(1-\theta)^r(-\theta)^x$$

The conditional distribution of X is given t is

$$h(x|t) = \binom{-r}{x}\binom{-nr + r}{t - x} \bigg/ \binom{-nr}{t} \quad (10.5)$$

These may be evaluated numerically quite easily, and a short table of the resulting fractiles is given in Appendix 11.

10.2 Outliers in time series

Fox (1972) considers the problem of outliers in a time series generated by an autoregressive model

$$U_t = \sum_{r=1}^{p} \alpha_r U_{t-r} + Z_t + \delta_t \quad (10.6)$$

where the Z_t are independently and identically distributed as $N(0, \sigma^2)$.

The series observed is X_t. Two types of outlier model are:

Type 1:

$$\begin{aligned} X_t &= U_t & t &\neq q \\ &= U_t + \Delta & t &= q \\ \delta_t &= 0 & &\textit{for all } t \end{aligned}$$

where q is unknown.

Type 2:

$$\begin{aligned} X_t &= U_t & &\text{for all } t \\ \delta_t &= 0 & t &\neq q \\ &= \Delta & t &= q \end{aligned} \tag{10.7}$$

The distinction between the two types is that in a type 1 outlier, the subsequent observations are not affected; only X_q is contaminated. In the type 2 model, some 'innovation' occurs at time q, and all subsequent observations are affected by it.

Under the type 1 outlier model, the X process under the null hypothesis $H_0 : \Delta = 0$, is a stationary autoregression with covariance matrix $\sigma^2 \mathbf{W}$, where $\mathbf{W}$ depends only on the autoregressive parameters. Fox points out that the $i, i \pm r$ element of $\mathbf{W}$ is asymptotically

$$\sum_{i=0}^{p-r} \alpha_i \alpha_{i+|r|} \tag{10.8}$$

and suggests the following procedure:

(i) Estimate the α_i and hence $\mathbf{W}^{-1}$.

(ii) Assuming q to be known, estimate $\mathbf{\Delta} = \Delta(0, 0, \ldots, 1, 0, \ldots, 0)$, the displacement of X_q.

(iii) Set up the likelihood ratio criterion

$$L_q = (\mathbf{X} - \hat{\Delta})' \hat{\mathbf{W}}^{-1} (\mathbf{X} - \hat{\Delta}) / \mathbf{X}' \hat{\mathbf{W}}^{-1} \mathbf{X}. \tag{10.9}$$

If $\mathbf{W}$ were known, L_q^{-1} would be $1 + F_{1,n-2}/(n-2)$, where $F_{1,n-2}$ represents an F-distributed variate with 1 and $n-2$ degrees of freedom.

Since in practice $\mathbf{W}$ must be estimated, this F-distribution is not applicable; however, by the consistency of the estimator of W, one would expect the F-distribution to provide a good approximation for n large. Fox verified, by the results of a simulation study, that this expectation is justified, even for fairly short series.

Since q is in general unknown, this result cannot be applied at once. To both estimate q and carry out the test, one finds L_q for each q, and selects that q for which

$$F_q = (n-2)(L_q^{-1} - 1)$$

TABLE 10.1

	α			
n	0.10	0.05	0.025	0.01
50	9.7	10.8	12.3	14.8
100	11.3	12.8	14.3	15.8
200	11.9	13.2	14.7	16.3
500	12.4	13.5	15.0	16.8

is a maximum. Fox studied the fractiles of this maximum by simulation, producing the results shown in Table 10.1.

Fox did not consider the use of the Bonferroni inequality, which leads to using the α/n fractile of $F_{1,n-2}$. Some comparisons of these fractiles with the simulated fractiles in the table show that the Bonferroni approximation is fairly good. For example, the $n = 50$, $\alpha = 0.1$ fractile has a Bonferroni significance of 0.15. Given that the fractile itself is approximate, having been estimated from a simulation, the agreement is good enough to encourage the use of the Bonferroni approximation outside the region of Fox's table.

Type 2 outliers are handled in a different way. Suppose first that q is known. Let $\hat{\alpha}_{r0}$ denote the estimate of α_r under the null hypothesis $\Delta = 0$, and $\hat{\alpha}_{r1}$ the estimate under the alternative. The likelihood ratio criterion reduces to

$$L_q = \left\{ \frac{\sum_t \left(X_t - \sum_r \hat{\alpha}_{r1} X_{t-r} - \hat{\delta}_t \right)^2}{\sum_t \left(X_t - \sum_r \hat{\alpha}_{r0} X_{t-r} \right)^2} \right\}^{(n-p)/2} \qquad (10.10)$$

An asymptotically equivalent statistic is the ratio $\hat{\Delta}/\hat{se}_{\hat{\Delta}}$ This is

$$l_q = (X_q - \Sigma_r \hat{\alpha}_{r1} X_{q-r}) / \{ \Sigma_t (X_t - \Sigma_r \hat{\alpha}_{r0} X_{t-r})^2 / (n-p) \}^{1/2} \qquad (10.11)$$

This has the asymptotic distribution

$$l_q^2 \sim \{1 + (n-p-1) F_{n-p-1,1}\}/(n-p) \qquad (10.12)$$

Once again, if q is unknown, one may choose q to maximize the resultant test statistic, though this extension is not actually made by Fox.

The simulation studies carried out show up a very important property of the tests for both type 1 and type 2 outliers – the null distributions of the test criteria are hardly affected by the values of the α_r.

It is also apparent that there can be a serious power loss if the series is treated erroneously as if it were a random sample; however, there is very little power loss if the α_i are replaced by their estimates in computing the test statistics. The latter property implies that one could assess the power of the procedures quite accurately with a rather minor adaptation of the power theory for random samples, which are simply time series with all $\alpha_i = 0$.

We should note that Fox's type 2 outlier model is related quite closely to Box and Tiao's (1975) technique of 'intervention analysis'. This technique is directed at locating and identifying instantaneous changes of regime in time series, of which general class a type 2 outlier is a particular member.

The Box–Tiao approach would test whether an outlier were present from the ratio $\hat{\Delta}/\hat{se}(\hat{\Delta})$, a criterion which is the same as that of Fox, though the method of estimation differs.

Bibliography

Anderson, T.W. (1959). *Introduction to Multivariate Statistical Analysis*, John Wiley and Sons, New York.

Anderson, T.W. (1971). *The Statistical Analysis of Time Series*, John Wiley and Sons, New York.

Anderson, T.W., and Darling, D.A. (1954). A test of goodness of fit, *Journal of the American Statistical Association*, **49**, 765–769.

Andrews, D. (1974). A robust method for multiple linear regression. *Technometrics*, **16**, 523–531.

Anscombe, F.J. (1960). Rejection of outliers, *Technometrics*, **2**, 123–147.

Anscombe, F.J. (1961). Examination of residuals, *Proceedings of the Fourth Berkeley Symposium in Probability and Mathematical Statistics*, 1–36.

Basu, A.P. (1965). On some tests of hypotheses relating to the exponential distribution when some outliers are present, *Journal of the American Statistical Association*, **60**, 548–559.

Basu, D. (1955). On statistics independent of a complete sufficient statistic, *Sankhyā*, **15**, 377–380.

Beckman, R.J., and Trussell, H.J. (1975). The distribution of an arbitrary studentized residual and the effect of updating in multiple regression, *Journal of the American Statistical Association*, **69**, 199–201.

Behnken, D.W., and Draper, N.R. (1972). Residuals and their variance patterns, *Technometrics*, **14**, 101–111.

Birnbaum, A. (1959). On the analysis of factorial experiments without replication, *Technometrics*, **1**, 343–357.

Bliss, C.I., Cochran, W.G., and Tukey, J.W. (1956). A rejection rule based upon the range, *Biometrika*, **43**, 418–422.

Bofinger, V.J. (1965). The *k* sample slippage problem, *Australian Journal of Statistics*, **7**, 20–31.

Borenius, G. (1959). On the distribution of the extreme values in a sample from a normal distribution, *Skandinavisk Aktuarietidskrift*, **1958**, 131–166.

Borenius, G. (1966). On the limit distribution of an extreme value in a sample from the normal distribution, *Skandinavisk Aktuarietidskrift*, **1965**, 1–15.

Box, G.E.P., and Tiao, G.C. (1968). A Bayesian approach to some outlier problems, *Biometrika*, **55**, 119–129.

Box, G.E.P., and Tiao, G.C. (1975). Intervention analysis with applications to economic and environmental problems, *Journal of the American Statistical Association*, **70**, 70–79.

Bradu, D. (1975). Search and Anova methods for interaction originating in a small set of outliers, *EDV in Medizin und Biologie*, **4**, 93–100.

Bradu, D., and Kass, G.V. (1977). Detecting outliers in multiple regression, *Report WISK*, **249**, C.S.I.R., Pretoria.

Brown, M.B. (1975). Exploring interaction effects in the analysis of variance, *Applied Statistics*, **24**, 288–298.

Chauvenet, W. (1876). *A Manual of Spherical and Practical Astronomy*, J.B. Lippincott, Philadelphia.

Cochran, W.G. (1941). The distribution of the largest of a set of estimated variances as a fraction of their total, *Annals of Eugenics*, **11**, 47–52.

Collett, D., and Lewis, T. (1976). The subjective nature of outlier rejection procedures, *Applied Statistics*, **25**, 228–237.

Conover, W.J. (1968). Two *k* sample slippage tests, *Journal of the American Statistical Association*, **63**, 614–626.

Daniel, C. (1960). Locating outliers in factorial experiments, *Technometrics*, **2**, 149–156.

Darwin, J.H. (1957). The differences between consecutive members of a series of random variables arranged in order of size, *Biometrika*, **44**, 211–218.

David, H.A. (1956). On the application to statistics of an elementary theorem in probability, *Biometrika*, **43**, 85–91. (Correction 449–451).

David, H.A. (1970). *Order Statistics*, John Wiley and Sons, New York.

David, H.A., and Paulson, A.S. (1965). The performance of several tests for outliers, *Biometrika*, **52**, 429–436.

de Finetti, B. (1961). The Bayesian approach to the rejection of outliers. *Proceedings of the Fourth Berkeley Symposium on Mathematical Statistics and Probability* (Ed. Neyman, J.), 199–210.

Dixon, W.J. (1950). Analysis of extreme values, *Annals of Mathematical Statistics*, **21**, 488–506.

Dixon, W.J. (1953). Processing data for outliers, *Biometrics*, **9**, 74–89.

Doornbos, R. (1956). Significance of the smallest of a set of estimated normal variances, *Statistica Neerlandica*, **10**, 117–126.

Doornbos, R. (1959). Statistische methoden voor het aanwijzen van uitbijters, *Statistica Neerlandica*, **13**, 453–462.

Doornbos, R. (1966). *Slippage Tests*, Mathematical Centre Tracts 15, Mathematisch Centrum, Amsterdam.

Draper, N.R., Guttman, I., and Kanemasu, H. (1971). The distribution of certain regression statistics, *Biometrika*, **58**, 295–298.

Dunn, O.J. (1959). Confidence intervals for the means of dependent normally distributed variables, *Journal of the American Statistical Association*, **54**, 613–621.

Durbin, J. (1973). Distribution theory for tests based on the sample distribution function, *SIAM*, Philadelphia.

Elashoff, J.D. (1972). A model for quadratic outliers in linear regression, *Journal of the American Statistical Association*, **67**, 478–485.

Ellenberg, J.H. (1973). The joint distribution of the standardized least squares residuals from a general linear regression, *Journal of the American Statistical Association*, **68**, 941–943.

Engelman, L., and Hartigan, J.A. (1969). Percentage points of a test for clusters, *Journal of the American Statistical Association*, **64**, 1647–1648.

Ferguson, T.S. (1961a). On the rejection of outliers, *Proceedings of the Fourth Berkeley Symposium on Mathematical Statistics and Probability*, 253–287.

Ferguson, T.S. (1961b). Rules for rejection of outliers, *Review of the International Statistical Institute*, **29**, 29–43.

Finney, D.J. (1941). The joint distribution of variance ratios based on a common mean square, *Annals of Eugenics*, **11**, 136–140.

Fisher, R.A. (1929). Tests of significance in harmonic analysis, *Proceedings of the Royal Society*, **A125**, 54–59.

Fox, A.J. (1972). Outliers in time series, *Journal of the Royal Statistical Society*, Series B, **34**, 350–363.

Fraser, D.A.S., and Guttman, I. (1956). Tolerance regions, *Annals of Mathematical Statistics*, **27**, 162–179.

Furnival, G.M. and Wilson, R.W. (1974). Regression by leaps and bounds, *Technometrics*, **16**, 499–511.

Gebhardt, F. (1964). On the risk of some strategies for outlying observations, *Annals of Mathematical Statistics*, **35**, 1524–1536.

Gentleman, J.F., and Wilk, M.B. (1975a). Detecting outliers in a two way table. I. Statistical behaviour of residuals, *Technometrics*, **17**, 1–14.

Gentleman, J.F., and Wilk, M.B. (1975b). Detecting outliers. II. Supplementing the direct analysis of residuals, *Biometrics*, **31**, 387–410.

Glaisher, J.W.L. (1872). On the law of facility of errors of observation and on the method of least squares, *Memoirs of the Royal Astronomical Society*, **39**, 75–124.

Gnanadesikan, R., and Kettenring, J.R. (1972). Robust estimates, residuals and outlier detection with multiresponse data, *Biometrics*, **28**, 81–124.

Godwin, H.J. (1945). On the distribution of the estimate of mean deviation obtained from samples from a normal population, *Biometrika*, **33**, 254–256.

Goodwin, H.M. (1913). *Elements of the Precision of Measurements and Graphical Methods*, McGraw-Hill, New York.

Granger, C.W.J., and Neave, H.R. (1968). A quick test for slippage, *Review of the International Statistical Institute*, **36**, 309–312.

Green, R.F. (1976). Outlier-prone and outlier-resistant distributions, *Journal of the American Statistical Association*, **71**, 502–505.

Grubbs, F.E. (1950). Sample criteria for testing outlying observations, *Annals of Mathematical Statistics*, **21**, 27–58.

Grubbs, F.E. (1969). Procedures for detecting outlying observations in samples, *Technometrics*, **11**, 1–21.

Grubbs, F.E., and Beck, G. (1972). Extension of sample sizes and percentage points for significance tests of outlying observations, *Technometrics*, **14**, 847–854.

Guttman, I. (1970). *Statistical Tolerance Regions: Classical and Bayesian*, Griffin, London.

Guttman, I. (1973). Premium and protection of several procedures for dealing with outliers when sample sizes are moderate to large, *Technometrics*, **15**, 385–404.

Guttman, I. (1973). Care and handling of univariate or multivariate outliers in detecting spuriosity—a Bayesian approach, *Technometrics*, **15**, 723–738.

Guttman, I., and Dutter, R. (1976). Procedures for investigating outliers when estimating in the general univariate linear situation—nonfull rank case, *Communications in Statistics*, **A5**, 819–836.

Guttman, I., and Smith, D.E. (1969). Investigation of rules for dealing with outliers in small samples from the normal distribution. I. Estimation of the mean, *Technometrics*, **11**, 527–554.

Guttman, I., and Smith, D.E. (1971). Investigation of rules for dealing with outliers in small samples from the normal distribution. II Estimation of the variance, *Technometrics*, **13**, 101–111.

Halperin, M., Greenhouse, S., Cornfield, J., and Zolokar, J. (1955). Tables of percentage points for the studentized maximum absolute deviation in normal samples, *Journal of the American Statistical Association*, **50**, 185–195.

Harter, H.L. (1974, 1975, 1976). The method of least squares and some alternatives. *International Statistical Review*, Pt I, **42**, 147–174; Pt II, **42**, 235–264; Pt III, **43**, 1–44; Pt IV, **43**, 125–190; Pt V, **43**, 269–278; Pt VI, **44**, 113–159.

Hartigan, J.A. (1972). Direct clustering of a data matrix, *Journal of the American Statistical Association*, **67**, 123–129.

Hartigan, J.A. (1975). *Clustering Algorithms*, John Wiley and Sons, New York.

Hartley, H.O. (1938). Studentization and large sample theory, *Supplement, Journal of the Royal Statistical Society*, **5**, 80–88.

Hawkins, D.M. (1969a). On the distribution and power of a test for a single outlier, *South African Statistical Journal*, **3**, 9–15.

Hawkins (1969b). Mathematical analysis of optimal outlier tests, PhD thesis, University of the Witwatersrand.

Hawkins, D.M. (1972). Analysis of a slippage test for the chisquared distribution, *South African Statistical Journal*, **6**, 11–17.

Hawkins, D.M. (1973). Repeated testing for outliers, *Statistica Neerlandica*, **27**, 1–10.

Hawkins, D.M. (1974). The detection of errors in multivariate data using principal components, *Journal of the American Statistical Association*, **69**, 340–344.

Hawkins, D.M. (1976). A general approach to outlier testing for the exponential family, *Institute of Statistics Mimeo Series*, no. 1090, Consolidated University of North Carolina.

Hawkins, D.M. (1978a). Analysis of three tests for one or two outliers, *Statistica Neerlandica*, **32**, 137–148.

Hawkins, D.M. (1978b). Fractiles of an extended multiple outlier test, *Journal of Statistical Computation and Simulation*, **8**, 227–236.

Hawkins, D.M., and Perold, A.F. (1977). On the joint distribution of left- and right-sided outlier statistics, *Utilitas Mathematica*, **12**, 129–143.

Heck, D.L. (1960). Charts of some upper percentage points of the distribution of the largest characteristic root, *Annals of Mathematical Statistics*, **31**, 625–642.

Irwin, J.O. (1925). On a criterion for the rejection of outlying observations, *Biometrika*, **17**, 238–250.

Ito, K. (1962). A comparison of the powers of two multivariate analysis of variance tests, *Biometrika*, **49**, 455–462.

John, J.A. and Prescott, P. (1975). Critical values of a test to detect outliers in factorial experiments, *Applied Statistics*, **24**, 56–59.

Johnson, N.L. and Kotz, S. (1969). *Discrete Distributions*, Houghton Mifflin, New York.

Johnson, N.L. and Welch, B.L. (1939). Some applications of the non-central *t* distribution, *Biometrika*, **31**, 362–389.

Joshi, P.C. (1972). Efficient estimation of the mean of an exponential distribution when an outlier is present, *Technometrics*, **14**, 137–143.

Joshi, S., and Sathe, Y.S. (1978). A generalization of Mosteller's test, *Communications in Statistics*, **A7**, 709–715.

Karlin, S., and Truax, D.R. (1960). Slippage problems, *Annals of Mathematical Statistics*, **31**, 296–324.

Kass, G.V. (1975). Significance testing in automatic interaction detection (AID), *Applied Statistics*, **24**, 178–189.

Kimball, A.E. (1951). On dependent tests of significance in analysis of variance, *Annals of Mathematical Statistics*, **22**, 600–602.

Kruskal, W.H. (1960). Some remarks on wild observations, *Technometrics*, **2**, 1–3.

Kudô, A. (1956). On the testing of outlying observations, *Sankhyā*, **17**, 67–76.

Kudô, A. (1957). The extreme value in a multivariate normal sample, *Memoirs*, Faculty of Science, Kyushu University A, **11**, 143–156.

Lehmann, E.L. (1959). *Testing Statistical Hypotheses*, John Wiley and Sons, New York.

Lehmann, E.L. (1975). *Nonparametrics: Statistical Methods based on Ranks*, McGraw-Hill, San Francisco.

Lingappaiah, G.S. (1976). Effects of outliers on the estimation of parameters, *Metrika*, **23**, 27–30.

Lund, R.E. (1975). Tables of an approximate test for outliers in linear models, *Technometrics*, **17**, 473–476.

McKay, A.T. (1935). The distribution of the difference between the extreme observation and the sample mean in samples of n from a normal universe, *Biometrika*, **27**, 466–471.

McMillan, R.G. (1971). Tests for one or two outliers in normal samples with unknown variance, *Technometrics*, **13**, 87–100.

McMillan, R.G., and David, H.A. (1971). Tests for one or two outliers in normal samples with known variance, *Technometrics*, **13**, 75–85.

Moran, M.A., and McMillan, R.G. (1973). Tests for one or two outliers in normal samples with unknown variance: a correction, *Technometrics*, **15**, 637–640.

Moses, L.E. (1978). Charts for finding upper percentage points of Student's t in the range .01 to .00001, *Communications in Statistics*, **B7**, 479–490.

Mosteller, F. (1948). A k sample slippage test for an extreme population, *Annals of Mathematical Statistics*, **19**, 58–65.

Mosteller, F., and Tukey, J.W. (1950). Significance levels for a k sample slippage test, *Annals of Mathematical Statistics*, **21**, 120–123.

Murphy, R.B. (1951). On tests for outlying observations, PhD thesis, Princeton University.

Nair, K.R. (1948). The distribution of the extreme deviate from the sample mean and its studentized forms, *Biometrika*, **35**, 118–144.

Neave, H.R. (1972). Some quick tests for slippage, *The Statistician*, **21**, 197–208.

Neave, H.R. (1973). A power study of some tests for slippage, *The Statistician*, **22**, 269–287.

Neave, H.R. (1975). A quick and simple technique for general slippage problems, *Journal of the American Statistical Association*, **70**, 721–726.

Nemenyi, P. (1963). Distribution-free multiple comparisons, PhD thesis, Princeton University.

Paulson, E. (1943). A note on tolerance limits, *Annals of Mathematical Statistics*, **14**, 90–93.

Paulson, E. (1952). An optimum solution to the k sample slippage problem for the normal distribution, *Annals of Mathematical Statistics*, **23**, 610–616.

Pearson, E.S., and Chandra Sekar, C. (1936). The efficiency of statistical tools and a criterion for the rejection of outlying observations, *Biometrika*, **28**, 308–320.

Pearson, E.S., and Hartley, H.O. (1954). *Biometrika Tables for Statisticians Volume I*, Cambridge University Press, Cambridge.

Peirce, W. (1852). Criterion for the rejection of doubtful observations, *Astronomical Journal*, **2**, 161–163.

Plackett, R.L. (1950). Some theorems in least squares. *Biometrika*, **37**, 149–157.

Prescott, P. (1975). An approximate test for outliers in linear models, *Technometrics*, **17**, 129–132.

Prescott, P. (1978). Examination of the behaviour of tests for outliers when more than one outlier is present, *Applied Statistics*, **27**, 10–25.

Press, S.J. (1972). *Applied Multivariate Analysis*, Holt, Rinehart and Winston, New York.

Pyke, R. (1965). Spacings, *Journal of the Royal Statistical Society*, Series B, **27**, 395–436.

Quesenberry, C.P., and David, H.A. (1961). Some tests for outliers, *Biometrika*, **48**, 379–390.

Ramachandran, K.V., and Khatri, C.G. (1957). On a decision procedure based on the Tukey statistic, *Annals of Mathematical Statistics*, **28**, 802–806.

Rao, C.R. (1964). The use and interpretation of principal component analysis in applied research, *Sankhyā*, **A26**, 329–358.

Relles, D.A., and Rogers, W.H. (1977). Statisticians are fairly robust estimators of location, *Journal of the American Statistical Association*, **72**, 107–111.

Rider, P.R. (1933). *Criteria for Rejection of Observations*, Washington University Studies, St. Louis.

Rosner, B. (1975). On the detection of many outliers, *Technometrics*, **17**, 221–227.

Savage, I.R. (1956). Contributions to the theory of rank order statistics—the two sample case, *Annals of Mathematical Statistics*, **27**, 590–615.

Scheffe, H. (1958). *The Analysis of Variance*, John Wiley and Sons, New York.

Scott, A.J., and Knott, M. (1976). An approximate test for use with AID, *Applied Statistics*, **25**, 103–106.

Scott, J.R. (1975). A regression assessment statistic, *Applied Statistics*, **24**, 42–45.

Sheesley, J.H. (1977). Tests for outlying observations, *Journal of Quality Technology*, **9**, 38–41.

Siotani, M. (1959). The extreme value of the generalized distances of the individual points in the multivariate normal sample, *Annals of the Institute of Statistical Mathematics of Tokyo*, **10**, 183–206.

Srikantan, K.S. (1961). Testing for the single outlier in a regression model, *Sankhyā*, **23**, 251–260.

Steel, R.G.D. (1960). A rank sum test for comparing all pairs of treatments, *Technometrics*, **2**, 197–207.

Stefansky, W. (1971). Rejecting outliers by maximum normed residual, *Annals of Mathematical Statistics*, **42**, 35–45.

Stefansky, W. (1972). Rejecting outliers in factorial designs, *Technometrics*, **14**, 469–479.

Stone, E.J. (1868). On the rejection of discordant observations, *Monthly Notices of the Royal Astronomical Society*, **28**, 165–168.

Thompson, W.R. (1935). On a criterion for the rejection of observations and the distributions of the ratio of the deviation to the sample standard deviation, *Annals of Mathematical Statistics*, **6**, 214–219.

Tiao, G.C., and Guttman, I. (1967). Analysis of outliers with adjusted residuals, *Technometrics*, **9**, 541–568.

Tietjen, G.L., and Moore, R.M. (1972). Some Grubbs-type statistics for the detection of several outliers, *Technometrics*, **14**, 583–597.

Tietjen, G.L., Moore, R.H., and Beckman, R.J. (1973). Testing for a single outlier in simple linear regression, *Technometrics*, **15**, 717–721.

Tiku, M.L. (1974). A new statistic for testing normality, *Communications in Statistics*, **3**, 223–232.

Tiku, M.L. (1975). A new statistic for testing suspected outliers, *Communications in Statistics*, **4**, 737–752.

Tiku, M.L., Rai, K., and Mead, E. (1976). A new statistic for testing exponentiality, *Communications in Statistics*, **3**, 485–493.

Truax, D.R. (1953). An optimum test for the variances of normal distributions, *Annals of Mathematical Statistics*, **24**, 669–673.

Tukey, J.W. (1949). Comparing individual means in the analysis of variance, *Biometrics*, **5**, 99–114.

Venables, W.N. (1974). Algorithm AS77—Null distribution of the largest root statistic, *Applied Statistics*, **23**, 458–465.

Walsh, J.E. (1950). Some nonparametric tests of whether the largest observations of a set are too large or too small, *Annals of Mathematical Statistics*, **21**, 583–592.

Walsh, J.E. (1953). Correction to 'some nonparametric tests of whether the largest observations of a set are too large or too small', *Annals of Mathematical Statistics*, **24**, 134–135.

Walsh, J.E. (1958). Large sample nonparametric rejection of outlying observations, *Annals of the Institute of Statistical Mathematics*, **10**, 223–232.

Walsh, J.E., and Kelleher, G.J. (1973). Nonparametric estimation of mean and variance when a few 'sample' values possibly outliers, *Annals of the Institute of Statistical Mathematics*, **25**, 87–90.

Wani, J.K., and Kabe, D.G. (1973). Distribution of linear functions of ordered rectangular variates, *Skandinavisk Aktuarietidskrift*, **56**, 58–60.

Wilks, S.S. (1963). Multivariate statistical outliers, *Sankhyā*, **25**, 407–426.

APPENDIX 1

Fractiles of B and B^* for normal samples

Given $X_1, X_2, \ldots, X_n \sim \mathrm{N}(\xi, \sigma^2)$ and $U \sim \sigma^2 \chi_\nu^2$ where ξ and σ are unknown and U is independent of the X_i; defining

$$\bar{X} = \sum_1^n X_i/n$$

$$S = \sum_1^n (X_i - \bar{X})^2 + U$$

the statistics are defined as

$$B = \max_{1 \leq i \leq n} (X_i - \bar{X})/\sqrt{S}$$

$$B^* = \max_{1 \leq i \leq n} |X_i - \bar{X}|/\sqrt{S}$$

By the symmetry of the normal distribution, the quantity

$$\max_{1 \leq i \leq n} (\bar{X} - X_i)/\sqrt{S}$$

which is suitable for testing for an outlier on the left of the sample, is distributed like B.

The fractiles tabulated are exact, and were taken from Hawkins and Perold (1977). For values outside the ranges tabulated, the Bonferroni inequality may be used to give an approximation which is only slightly conservative for small α. Let

$$B_0 = t[(n-1)/\{n(n+\nu-2+t^2)\}]^{1/2}$$

where t is the upper α/n fractile of a t-variate with $n+\nu-2$ degrees of random. Then B_0 approximates the upper α fractile of B and the upper 2α fractile of B^*.

A convenient source for the extreme fractiles of the t-distribution needed for the approximation is Moses (1978).

Critical values of B^*

ν	0				5				15				30			
n	α 0.10	0.05	0.01	0.001	0.10	0.05	0.01	0.001	0.10	0.05	0.01	0.001	0.10	0.05	0.01	0.001
5	0.8357	0.8575	0.8818	0.8917	0.6396	0.6839	0.7573	0.8190	0.4583	0.5011	0.5796	0.6606	0.3468	0.3827	0.4510	0.5264
6	0.8119	0.8440	0.8823	0.9032	0.6380	0.6813	0.7554	0.8207	0.4690	0.5103	0.5869	0.6674	0.3593	0.3942	0.4612	0.5360
7	0.7912	0.8246	0.8733	0.9051	0.6325	0.6750	0.7493	0.8173	0.4749	0.5148	0.5898	0.6695	0.3678	0.4018	0.4676	0.5417
8	0.7679	0.8038	0.8596	0.9006	0.6251	0.6667	0.7409	0.8109	0.4777	0.5165	0.5901	0.6690	0.3736	0.4069	0.4715	0.5448
9	0.7458	0.7831	0.8439	0.8923	0.6168	0.6576	0.7314	0.8029	0.4787	0.5165	0.5886	0.6668	0.3777	0.4102	0.4738	0.5463
10	0.7254	0.7633	0.8274	0.8817	0.6081	0.6480	0.7213	0.7939	0.4784	0.5152	0.5861	0.6635	0.3805	0.4124	0.4750	0.5466
11	0.7064	0.7445	0.8108	0.8697	0.5993	0.6385	0.7111	0.7845	0.4771	0.5131	0.5828	0.6595	0.3824	0.4137	0.4753	0.5461
12	0.6889	0.7271	0.7947	0.8571	0.5906	0.6290	0.7010	0.7748	0.4752	0.5105	0.5790	0.6550	0.3836	0.4143	0.4750	0.5451
13	0.5728	0.7107	0.7791	0.8442	0.5820	0.6197	0.6910	0.7651	0.4729	0.5075	0.5749	0.6501	0.3842	0.4144	0.4742	0.5436
14	0.6578	0.6954	0.7642	0.8313	0.5737	0.6107	0.6812	0.7554	0.4703	0.5042	0.5706	0.6451	0.3843	0.4140	0.4731	0.5417
15	0.6438	0.6811	0.7500	0.8186	0.5656	0.6020	0.6717	0.7459	0.4674	0.5007	0.5662	0.6400	0.3842	0.4134	0.4717	0.5397
16	0.6308	0.6676	0.7364	0.8062	0.5578	0.5936	0.6625	0.7365	0.4644	0.4971	0.5617	0.6348	0.3837	0.4125	0.4701	0.5374
17	0.6187	0.6550	0.7235	0.7942	0.5503	0.5855	0.6535	0.7274	0.4612	0.4935	0.5571	0.6295	0.3831	0.4115	0.4683	0.5350
18	0.6073	0.6431	0.7112	0.7825	0.5430	0.5777	0.6449	0.7185	0.4580	0.4898	0.5526	0.6243	0.3822	0.4103	0.4665	0.5325
19	0.5965	0.6319	0.6996	0.7711	0.5361	0.5701	0.6365	0.7099	0.4548	0.4860	0.5480	0.6192	0.3812	0.4089	0.4645	0.5299
20	0.5864	0.6213	0.6884	0.7602	0.5293	0.5629	0.6286	0.7015	0.4516	0.4823	0.5436	0.6141	0.3801	0.4075	0.4624	0.5272
21	0.5769	0.6113	0.6778	0.7497	0.5229	0.5559	0.6209	0.6934	0.4483	0.4786	0.5392	0.6090	0.3789	0.4059	0.4603	0.5245
22	0.5679	0.6018	0.6677	0.7396	0.5166	0.5492	0.6134	0.6855	0.4451	0.4750	0.5348	0.6040	0.3776	0.4043	0.4581	0.5218
23	0.5593	0.5927	0.6581	0.7298	0.5106	0.5427	0.6062	0.6778	0.4419	0.4714	0.5305	0.5991	0.3763	0.4027	0.4559	0.5191
24	0.5512	0.5841	0.6488	0.7204	0.5048	0.5365	0.5993	0.6704	0.4387	0.4678	0.5263	0.5943	0.3749	0.4010	0.4537	0.5163
25	0.5434	0.5760	0.6400	0.7113	0.4992	0.5305	0.5925	0.6632	0.4356	0.4643	0.5221	0.5896	0.3735	0.3993	0.4515	0.5136
26	0.5360	0.5681	0.6315	0.7025	0.4938	0.5247	0.5861	0.6562	0.4325	0.4609	0.5180	0.5850	0.3720	0.3975	0.4492	0.5109
27	0.5289	0.5607	0.6234	0.6940	0.4886	0.5190	0.5798	0.6495	0.4294	0.4575	0.5140	0.5804	0.3705	0.3958	0.4470	0.5081
28	0.5222	0.5535	0.6156	0.6859	0.4836	0.5136	0.5737	0.6429	0.4264	0.4541	0.5101	0.5760	0.3690	0.3940	0.4447	0.5054
29	0.5157	0.5466	0.6081	0.6780	0.4787	0.5084	0.5678	0.6365	0.4234	0.4509	0.5063	0.5716	0.3675	0.3922	0.4425	0.5027
30	0.5095	0.5400	0.6009	0.6704	0.4740	0.5033	0.5621	0.6303	0.4205	0.4476	0.5025	0.5673	0.3660	0.3904	0.4403	0.5001

Critical values of B

n	ν 0				5				15				30			
	α 0.10	0.05	0.01	0.001	0.10	0.05	0.01	0.001	0.10	0.05	0.01	0.001	0.10	0.05	0.01	0.001
5	0.8008	0.8357	0.8744	0.8901	0.5855	0.6399	0.7297	0.8042	0.4137	0.4611	0.5489	0.6391	0.3119	0.3502	0.4242	0.5058
6	0.7732	0.8149	0.8695	0.8992	0.5857	0.6382	0.7273	0.8047	0.4251	0.4709	0.5567	0.6459	0.3243	0.3619	0.4346	0.5155
7	0.7463	0.7912	0.8562	0.8984	0.5820	0.6327	0.7207	0.8003	0.4319	0.4763	0.5601	0.6482	0.3331	0.3699	0.4413	0.5213
8	0.7215	0.7679	0.8394	0.8915	0.5762	0.6253	0.7122	0.7932	0.4358	0.4789	0.5608	0.6478	0.3394	0.3754	0.4456	0.5246
9	0.6991	0.7458	0.8214	0.8810	0.5693	0.6170	0.7026	0.7846	0.4377	0.4797	0.5598	0.6457	0.3440	0.3792	0.4482	0.5262
10	0.6787	0.7254	0.8032	0.8686	0.5619	0.6082	0.6926	0.7752	0.4383	0.4792	0.5577	0.6426	0.3473	0.3819	0.4497	0.5268
11	0.6603	0.7064	0.7856	0.8553	0.5543	0.5994	0.6825	0.7654	0.4379	0.4779	0.5548	0.6387	0.3497	0.3836	0.4504	0.5265
12	0.6435	0.6889	0.7687	0.8415	0.5466	0.5907	0.6725	0.7554	0.4368	0.4759	0.5515	0.6343	0.3513	0.3847	0.4504	0.5256
13	0.6280	0.6728	0.7526	0.8277	0.5391	0.5821	0.6627	0.7455	0.4353	0.4736	0.5477	0.6296	0.3524	0.3852	0.4500	0.5243
14	0.6138	0.6578	0.7373	0.8141	0.5318	0.5738	0.6531	0.7357	0.4334	0.4709	0.5438	0.6247	0.3530	0.3853	0.4491	0.5226
15	0.6007	0.6438	0.7229	0.8009	0.5246	0.5658	0.6439	0.7261	0.4312	0.4680	0.5397	0.6197	0.3533	0.3850	0.4480	0.5207
16	0.5885	0.6308	0.7093	0.7880	0.5177	0.5580	0.6349	0.7167	0.4288	0.4649	0.5355	0.6147	0.3533	0.3846	0.4467	0.5186
17	0.5771	0.6187	0.6964	0.7756	0.5109	0.5505	0.6262	0.7075	0.4263	0.4618	0.5313	0.6096	0.3530	0.3839	0.4452	0.5164
18	0.5665	0.6073	0.6841	0.7636	0.5045	0.5432	0.6179	0.6987	0.4237	0.4586	0.5271	0.6046	0.3526	0.3830	0.4436	0.5140
19	0.5565	0.5966	0.6726	0.7521	0.4982	0.5363	0.6099	0.6900	0.4210	0.4553	0.5229	0.5995	0.3520	0.3820	0.4418	0.5116
20	0.5471	0.5865	0.6616	0.7410	0.4922	0.5296	0.6021	0.6817	0.4183	0.4521	0.5187	0.5946	0.3512	0.3808	0.4400	0.5091
21	0.5383	0.5770	0.6512	0.7304	0.4863	0.5231	0.5947	0.6736	0.4156	0.4488	0.5146	0.5897	0.3504	0.3796	0.4381	0.5065
22	0.5299	0.5680	0.6412	0.7202	0.4807	0.5169	0.5875	0.6658	0.4128	0.4456	0.5105	0.5848	0.3495	0.3783	0.4361	0.5040
23	0.5220	0.5594	0.6318	0.7104	0.4753	0.5109	0.5805	0.6582	0.4101	0.4423	0.5064	0.5801	0.3484	0.3770	0.4342	0.5014
24	0.5145	0.5513	0.6228	0.7009	0.4701	0.5051	0.5738	0.6509	0.4073	0.4391	0.5025	0.5754	0.3474	0.3755	0.4321	0.4987
25	0.5073	0.5435	0.6141	0.6919	0.4650	0.4995	0.5674	0.6438	0.4046	0.4360	0.4986	0.5708	0.3462	0.3741	0.4301	0.4961
26	0.5005	0.5361	0.6059	0.6831	0.4601	0.4941	0.5611	0.6369	0.4019	0.4329	0.4947	0.5663	0.3451	0.3726	0.4281	0.4935
27	0.4939	0.5291	0.5980	0.6747	0.4554	0.4889	0.5551	0.6302	0.3992	0.4298	0.4910	0.5619	0.3438	0.3711	0.4260	0.4909
28	0.4877	0.5224	0.5904	0.6666	0.4508	0.4838	0.5493	0.6238	0.3966	0.4268	0.4873	0.5576	0.3426	0.3696	0.4240	0.4884
29	0.4818	0.5159	0.5832	0.6588	0.4464	0.4790	0.5436	0.6175	0.3940	0.4239	0.4836	0.5534	0.3414	0.3680	0.4219	0.4858
30	0.4760	0.5097	0.5762	0.6513	0.4421	0.4743	0.5382	0.6114	0.3915	0.4209	0.4801	0.5492	0.3401	0.3665	0.4199	0.4832

APPENDIX 2

Fractiles of L_k for normal samples

Given $X_1, X_2, \ldots, X_n \sim N(\xi, \sigma^2)$, let $X_{(1)} \leqslant X_{(2)} \leqslant \ldots \leqslant X_{(n)}$ be the corresponding order statistics. Define

$$\bar{X}_{0,k} = \sum_{i=1}^{n-k} X_{(i)}/(n-k)$$

$$S^2_{0,k} = \sum_{i=1}^{n-k} (X_{(i)} - \bar{X}_{0,k})^2$$

then low values of $L_k = S^2_{0,k}/S^2_{0,0}$ are used for testing for the presence of k outliers on the right. By symmetry, an analogous test may be used to test for k outliers on the left.

The fractiles tabulated were estimated from simulation of 20 000 random samples.

No really satisfactory general approximation is known for values of n and k outside the range simulated. Some additional values for the case $k = 2$ are given in Appendix 5. The fractiles of E_k given in Appendix 3 provide conservative bounds for the fractiles of L_k, but these bounds become increasingly and excessively conservative for large k.

$\alpha = 0.100$

					k					
n	1	2	3	4	5	6	7	8	9	10
5	0.1963	0.0389								
6	0.2823	0.0931	0.0205							
7	0.3481	0.1476	0.0566							
8	0.4024	0.1974	0.0938	0.0375						
9	0.4517	0.2466	0.1339	0.0679						
10	0.4889	0.2873	0.1700	0.0974	0.0503					
15	0.6176	0.4360	0.3204	0.2355	0.1719	0.1238	0.0862			
20	0.6865	0.5265	0.4174	0.3360	0.2729	0.2198	0.1763	0.1407	0.1100	0.0843
25	0.7322	0.5907	0.4909	0.4129	0.3482	0.2956	0.2514	0.2129	0.1796	0.1502
30	0.7662	0.6378	0.5465	0.4726	0.4109	0.3597	0.3156	0.2764	0.2417	0.2111
35	0.7923	0.6754	0.5888	0.5188	0.4611	0.4112	0.3678	0.3290	0.2942	0.2629
40	0.8112	0.7026	0.6207	0.5549	0.5004	0.4536	0.4112	0.3738	0.3398	0.3086
45	0.8281	0.7269	0.6504	0.5881	0.5349	0.4889	0.4471	0.4106	0.3772	0.3469
50	0.8403	0.7453	0.6739	0.6145	0.5641	0.5199	0.4801	0.4442	0.4113	0.3813
60	0.8615	0.7769	0.7123	0.6582	0.6110	0.5696	0.5325	0.4986	0.4679	0.4392
70	0.8773	0.8019	0.7426	0.6932	0.6501	0.6107	0.5756	0.5439	0.5144	0.4871
80	0.8893	0.8201	0.7650	0.7186	0.6782	0.6417	0.6091	0.5790	0.5514	0.5251
90	0.8991	0.8350	0.7834	0.7400	0.7018	0.6676	0.6363	0.6078	0.5812	0.5564
100	0.9074	0.8476	0.7997	0.7589	0.7228	0.6902	0.6607	0.6331	0.6080	0.5845

$\alpha = 0.050$

					k					
n	1	2	3	4	5	6	7	8	9	10
5	0.1239	0.0187								
6	0.1971	0.0561	0.0096							
7	0.2656	0.1025	0.0341							
8	0.3195	0.1446	0.0635	0.0218						
9	0.3772	0.1922	0.0985	0.0460						
10	0.4187	0.2297	0.1311	0.0706	0.0331					
15	0.5613	0.3850	0.2742	0.1964	0.1398	0.0975	0.0658			
20	0.6381	0.4812	0.3761	0.2971	0.2366	0.1889	0.1483	0.1153	0.0886	0.0666
25	0.6916	0.5492	0.4500	0.3746	0.3129	0.2627	0.2206	0.1852	0.1545	0.1270
30	0.7321	0.6013	0.5098	0.4381	0.3782	0.3265	0.2835	0.2457	0.2132	0.1848
35	0.7615	0.6428	0.5544	0.4848	0.4277	0.3790	0.3363	0.2990	0.2666	0.2362
40	0.7855	0.6721	0.5901	0.5256	0.4701	0.4224	0.3813	0.3441	0.3119	0.2821
45	0.8035	0.7000	0.6220	0.5584	0.5050	0.4594	0.4195	0.3834	0.3513	0.3223
50	0.8181	0.7209	0.6479	0.5878	0.5366	0.4911	0.4516	0.4160	0.3850	0.3565
60	0.8428	0.7553	0.6884	0.6335	0.5861	0.5448	0.5074	0.4743	0.4434	0.4154
70	0.8610	0.7833	0.7222	0.6715	0.6276	0.5892	0.5536	0.5215	0.4918	0.4642
80	0.8746	0.8021	0.7453	0.6980	0.6576	0.6209	0.5873	0.5569	0.5291	0.5033
90	0.8863	0.8195	0.7668	0.7227	0.6834	0.6486	0.6163	0.5880	0.5611	0.5365
100	0.8957	0.8333	0.7844	0.7419	0.7057	0.6729	0.6432	0.6150	0.5898	0.5656

$\alpha = .025$

	k									
n	1	2	3	4	5	6	7	8	9	10
5	0.0789	0.0090								
6	0.1411	0.0346	0.0045							
7	0.2010	0.0729	0.0217							
8	0.2590	0.1092	0.0436	0.0137						
9	0.3158	0.1509	0.0729	0.0320						
10	0.3553	0.1886	0.1020	0.0520	0.0221					
15	0.5063	0.3389	0.2371	0.1671	0.1155	0.0781	0.0506			
20	0.5950	0.4388	0.3384	0.2647	0.2065	0.1635	0.1270	0.0966	0.0725	0.0535
25	0.6557	0.5133	0.4172	0.3401	0.2836	0.2348	0.1967	0.1633	0.1345	0.1102
30	0.7004	0.5688	0.4780	0.4068	0.3489	0.2990	0.2579	0.2232	0.1908	0.1639
35	0.7314	0.6141	0.5238	0.4560	0.3985	0.3499	0.3106	0.2740	0.2425	0.2141
40	0.7566	0.6443	0.5637	0.4967	0.4429	0.3970	0.3559	0.3203	0.2890	0.2602
45	0.7787	0.6741	0.5963	0.5342	0.4812	0.4358	0.3963	0.3603	0.3289	0.2992
50	0.7978	0.6990	0.6254	0.5643	0.5125	0.4679	0.4295	0.3947	0.3633	0.3347
60	0.8244	0.7344	0.6651	0.6107	0.5626	0.5216	0.4857	0.4525	0.4220	0.3938
70	0.8451	0.7652	0.7039	0.6519	0.6063	0.5680	0.5330	0.5009	0.4713	0.4449
80	0.8613	0.7857	0.7269	0.6792	0.6380	0.6018	0.5692	0.5389	0.5115	0.4851
90	0.8730	0.8054	0.7523	0.7059	0.6665	0.6318	0.6002	0.5709	0.5443	0.5192
100	0.8846	0.8203	0.7699	0.7274	0.6902	0.6569	0.6263	0.5986	0.5729	0.5492

$\alpha = .010$

					k					
n	1	2	3	4	5	6	7	8	9	10
5	0.0425	0.0036								
6	0.0920	0.0192	0.0017							
7	0.1434	0.0453	0.0108							
8	0.1947	0.0756	0.0252	0.0078						
9	0.2429	0.1102	0.0496	0.0190						
10	0.2842	0.1417	0.0738	0.0363	0.0138					
15	0.4422	0.2908	0.1981	0.1354	0.0911	0.0599	0.0375			
20	0.5432	0.3924	0.2968	0.2268	0.1744	0.1341	0.1030	0.0781	0.0576	0.0406
25	0.6139	0.4717	0.3784	0.3046	0.2509	0.2063	0.1694	0.1392	0.1125	0.0895
30	0.6611	0.5283	0.4415	0.3706	0.3148	0.2674	0.2280	0.1963	0.1673	0.1420
35	0.6983	0.5736	0.4899	0.4231	0.3656	0.3197	0.2817	0.2484	0.2173	0.1903
40	0.7245	0.6134	0.5302	0.4642	0.4109	0.3654	0.3269	0.2926	0.2617	0.2344
45	0.7494	0.6456	0.5648	0.5049	0.4528	0.4056	0.3668	0.3315	0.3005	0.2729
50	0.7709	0.6700	0.5964	0.5353	0.4860	0.4418	0.4034	0.3690	0.3382	0.3106
60	0.8019	0.7105	0.6402	0.5839	0.5377	0.4975	0.4606	0.4251	0.3954	0.3691
70	0.8231	0.7414	0.6806	0.6290	0.5851	0.5455	0.5108	0.4790	0.4500	0.4232
80	0.8432	0.7654	0.7082	0.6583	0.6169	0.5801	0.5472	0.5171	0.4897	0.4640
90	0.8564	0.7879	0.7331	0.6878	0.6489	0.6121	0.5804	0.5506	0.5240	0.4994
100	0.8694	0.8026	0.7519	0.7085	0.6706	0.6369	0.6075	0.5792	0.5539	0.5292

APPENDIX 3

Fractiles of E_k for normal samples

Given $X_1, X_2, \ldots, X_n \sim N(\xi, \sigma^2)$ and independently $U \sim \sigma^2\chi^2_\nu$:
Let

$$\bar{X} = \sum_1^n X_i/n, \; S = \sum_1^n (X_i - \bar{X})^2 + U$$

Using the procedure set out in Section 5.1.2, sort the X_i into k outliers and $n - k$ inliers, renumbering the latter $X_{k+1}, \ldots, X_n$. Define

$$\bar{X}_k = \sum_{k+1}^n X_i/(n-k), \quad S_k = \sum_{k+1}^n (X_i - \bar{X}_k)^2 + U, \quad E_k = S_k/S$$

Small values of E_k indicate that the k observations put in the 'outlier' partition are in fact contaminants from a different distribution.

The fractiles are taken from Hawkins (1978b). They were obtained by fitting a mathematical function defined below to the results of 130 simulation runs in each of which 2000 random samples were drawn.

For values of n, ν and α not covered in Table A3.1, a more tedious computation using Table A3.2 is required.
Define

$$\begin{aligned} c &= h_{4k}\{\exp(-0.05\nu) - 0.5953\} \\ d &= h_{2k} + h_{1k}\log_e\{n(n-1)\ldots(n-k+1)\} \\ f &= h_{3k}\{\log_e \alpha + 3.778\} \\ a &= d + c + h_{5k}dc \\ b &= f + c + h_{6k}fc \\ \alpha_0 &= \exp(h_{0k} + a + b) \end{aligned}$$

The required fractile E_0 of E_k is then the α_0 fractile of a beta type 1 distribution with $\frac{1}{2}(n + \nu - k - 1)$ and $\frac{1}{2}k$ degrees of freedom, i.e.

$$\int_0^{E_0} r^{1/2(n+\nu-k-1)-1}(1-r)^{1/2k-1}\,dr = \alpha_0 B\{\tfrac{1}{2}(n+\nu-k-1), \tfrac{1}{2}k\}$$

TABLE A3.1 Fractiles of E_k

	ν		0			10			20	
k	n	α 0.05	0.01	0.001	0.05	0.01	0.001	0.05	0.01	0.001
2	5	0.0061	0.0011	0.0001	0.4556	0.3380	0.2205	0.6613	0.5610	0.4433
2	10	0.1640	0.0995	0.0487	0.4920	0.3985	0.2948	0.6461	0.5650	0.4664
2	15	0.3104	0.2319	0.1529	0.5404	0.4592	0.3638	0.6596	0.5890	0.5010
2	20	0.4136	0.3367	0.2509	0.5829	0.5105	0.4222	0.6780	0.6148	0.5345
2	25	0.4886	0.4168	0.3320	0.6186	0.5531	0.4712	0.6964	0.6388	0.5647
2	30	0.5455	0.4792	0.3982	0.6486	0.5887	0.5126	0.7135	0.6606	0.5917
2	40	0.6262	0.5698	0.4977	0.6959	0.6449	0.5782	0.7432	0.6975	0.6369
2	50	0.6810	0.6322	0.5684	0.7314	0.6869	0.6278	0.7676	0.7272	0.6731
2	75	0.7639	0.7277	0.6788	0.7907	0.7569	0.7111	0.8117	0.7804	0.7377
2	100	0.8109	0.7821	0.7428	0.8275	0.8002	0.7628	0.8412	0.8156	0.7803
3	10	0.0740	0.0395	0.0160	0.4000	0.3117	0.2190	0.5800	0.4959	0.3973
3	15	0.1967	0.1389	0.0847	0.4454	0.3684	0.2814	0.5860	0.5141	0.4270
3	20	0.2972	0.2338	0.1661	0.4892	0.4197	0.3376	0.6028	0.5386	0.4591
3	25	0.3758	0.3128	0.2410	0.5278	0.4641	0.3867	0.6217	0.5632	0.4896
3	30	0.4379	0.3776	0.3057	0.5612	0.5024	0.4292	0.6402	0.5863	0.5175
3	40	0.5297	0.4758	0.4084	0.6153	0.5642	0.4988	0.6737	0.6268	0.5657
3	50	0.5942	0.5463	0.4846	0.6570	0.6119	0.5530	0.7021	0.6604	0.6053
3	75	0.6948	0.6579	0.6087	0.7286	0.6936	0.6467	0.7551	0.7223	0.6782
3	100	0.7534	0.7236	0.6832	0.7741	0.7456	0.7067	0.7915	0.7645	0.7275
4	10	0.0299	0.0132	0.0041	0.3360	0.2515	0.1673	0.5357	0.4478	0.3485
4	15	0.1239	0.0819	0.0455	0.3753	0.3022	0.2225	0.5313	0.4583	0.3721
4	20	0.2154	0.1631	0.1099	0.4182	0.3516	0.2752	0.5449	0.4803	0.4019
4	25	0.2923	0.2370	0.1760	0.4575	0.3959	0.3227	0.5631	0.5043	0.4316
4	30	0.3558	0.3006	0.2368	0.4924	0.4350	0.3650	0.5819	0.5277	0.4595
4	40	0.4530	0.4015	0.3383	0.5504	0.4998	0.4360	0.6174	0.5700	0.5080
4	50	0.5235	0.4764	0.4169	0.5962	0.5509	0.4926	0.6483	0.6059	0.5506
4	75	0.6366	0.5992	0.5497	0.6765	0.6408	0.5933	0.7076	0.6739	0.6289
4	100	0.7041	0.6733	0.6319	0.7287	0.6992	0.6593	0.7492	0.7212	0.6832
5	10	0.0096	0.0033	0.0007	0.2906	0.2083	0.1308	0.5072	0.4139	0.3123
5	15	0.0763	0.0470	0.0236	0.3214	0.2518	0.1786	0.4891	0.4146	0.3290
5	20	0.1560	0.1138	0.0727	0.3618	0.2982	0.2271	0.4984	0.4333	0.3559
5	25	0.2283	0.1806	0.1296	0.4006	0.3413	0.2723	0.5151	0.4561	0.3843
5	30	0.2906	0.2412	0.1852	0.4359	0.3803	0.3136	0.5336	0.4792	0.4117
5	40	0.3895	0.3412	0.2828	0.4960	0.4463	0.3844	0.5699	0.5222	0.4615
5	50	0.4634	0.4182	0.3616	0.5443	0.4994	0.4421	0.6024	0.5596	0.5042
5	75	0.5854	0.5482	0.4994	0.6310	0.5950	0.5473	0.6662	0.6319	0.5863
5	100	0.6600	0.6289	0.5872	0.6885	0.6584	0.6180	0.7120	0.6832	0.6444
6	15	0.0451	0.0254	0.0112	0.2790	0.2121	0.1445	0.4562	0.3798	0.2943
6	20	0.1123	0.0784	0.0472	0.3160	0.2550	0.1887	0.4602	0.3945	0.3181
6	25	0.1786	0.1375	0.0949	0.3535	0.2964	0.2313	0.4748	0.4156	0.3448
6	30	0.2382	0.1938	0.1448	0.3886	0.3346	0.2710	0.4925	0.4381	0.3715
6	40	0.3964	0.2910	0.2371	0.4495	0.4007	0.3407	0.5289	0.4811	0.4209
6	50	0.4121	0.3685	0.3146	0.4995	0.4550	0.3988	0.5624	0.5193	0.4641
6	75	0.5403	0.5033	0.4551	0.5909	0.5546	0.5070	0.6296	0.5948	0.5488
6	100	0.6205	0.5890	0.5471	0.6526	0.6220	0.5811	0.6787	0.6493	0.6099
7	15	0.0249	0.0127	0.0049	0.2463	0.1821	0.1194	0.4331	0.3543	0.2685
7	20	0.0794	0.0534	0.0304	0.2786	0.2207	0.1592	0.4295	0.3636	0.2882
7	25	0.1389	0.1045	0.0700	0.3141	0.2597	0.1989	0.4410	0.3822	0.3128

(Contd.)

TABLE A3.1 (*Contd.*)

	ν	0			10			20		
k	n	α 0.05	0.01	0.001	0.05	0.01	0.001	0.05	0.01	0.001
7	30	0.1949	0.1562	0.1143	0.3481	0.2966	0.2367	0.4573	0.4033	0.3381
7	40	0.2907	0.2493	0.2007	0.4088	0.3617	0.3044	0.4929	0.4425	0.3864
7	50	0.3668	0.3261	0.2761	0.4596	0.4163	0.3620	0.5266	0.4839	0.4294
7	75	0.4992	0.4635	0.4172	0.5542	0.5184	0.4716	0.5960	0.5612	0.5155
7	100	0.5838	0.5530	0.5120	0.6192	0.5888	0.5481	0.6477	0.6182	0.5786
8	20	0.0546	0.0347	0.0182	0.2474	0.1913	0.1335	0.4047	0.3373	0.2618
8	25	0.1070	0.0780	0.0498	0.2805	0.2279	0.1703	0.4125	0.3530	0.2841
8	30	0.1591	0.1247	0.0884	0.3133	0.2632	0.2060	0.4270	0.3727	0.3081
8	40	0.2517	0.2130	0.1682	0.3732	0.3270	0.2715	0.4611	0.4136	0.3550
8	50	0.3275	0.2884	0.2409	0.4243	0.3815	0.3284	0.4947	0.4518	0.3977
8	75	0.4628	0.4273	0.3817	0.5212	0.4853	0.4387	0.5656	0.5305	0.4849
8	100	0.5510	0.5199	0.4789	0.5889	0.5581	0.5172	0.6193	0.5893	0.5494
9	20	0.0365	0.0217	0.0104	0.2210	0.1666	0.1123	0.3838	0.3146	0.2391
9	25	0.0818	0.0576	0.0351	0.2513	0.2005	0.1462	0.3874	0.3274	0.2591
9	30	0.1294	0.0992	0.0682	0.2827	0.2342	0.1798	0.3999	0.3455	0.2816
9	40	0.2178	0.1820	0.1413	0.3414	0.2963	0.2428	0.4324	0.3850	0.3270
9	50	0.2924	0.2553	0.2107	0.3924	0.3503	0.2986	0.4658	0.4228	0.3691
9	75	0.4290	0.3943	0.3499	0.4909	0.4551	0.4089	0.5377	0.5023	0.4564
9	100	0.5200	0.4891	0.4485	0.5608	0.5298	0.4888	0.5931	0.5627	0.5225
10	20	0.0230	0.0126	0.0054	0.1982	0.1455	0.0947	0.3671	0.2961	0.2204
10	25	0.0611	0.0414	0.0239	0.2258	0.1769	0.1258	0.3658	0.3055	0.2379
10	30	0.1043	0.0780	0.0518	0.2557	0.2089	0.1574	0.3761	0.3218	0.2589
10	40	0.1881	0.1551	0.1181	0.3130	0.2692	0.2178	0.4068	0.3596	0.3026
10	50	0.2612	0.2260	0.1842	0.3636	0.3225	0.2723	0.4396	0.3970	0.3439
10	75	0.3986	0.3645	0.3212	0.4632	0.4277	0.3822	0.5121	0.4768	0.4311
10	100	0.4920	0.4612	0.4209	0.5349	0.5039	0.4631	0.5688	0.5384	0.4982

TABLE A3.2

k	h_{0k}	h_{1k}	h_{2k}	h_{3k}	h_{4k}	h_{5k}	h_{6k}
2	−8.3656	−0.8691	5.3583	1.1113	−0.1655	0.6914	0.2598
3	−9.3908	−0.8023	7.3391	1.2010	−0.3892	0.5329	0.1250
4	−10.4972	−0.7556	9.4674	1.2857	−0.6128	0.4290	0.0580
5	−11.5379	−0.7228	11.6167	1.3613	−0.8190	0.3521	0.0469
6	−12.5597	−0.6966	13.7865	1.4357	−1.0070	0.3015	0.0344
7	−13.5926	−0.6811	16.1551	1.4837	−1.1840	0.2700	0.0417
8	−14.7164	−0.6659	18.5844	1.5572	−1.3462	0.2561	0.0317
9	−15.9890	−0.6553	21.1983	1.6225	−1.4542	0.2420	0.0317
10	−17.4906	−0.6392	23.9625	1.6807	−1.5212	0.2437	0.0245

APPENDIX 4

Fractiles of $T_{n:i}$ for normal samples

Given $X_1, \ldots, X_n \sim N(\xi, \sigma^2)$ and independently $U \sim \sigma^2\chi^2_\nu$; re-order the $\{X_i\}$ so that X_i is the ith most aberrant observation in the sample. This ordering is done using the procedure set out in Section 5.1.2. The quantity K is prespecified, and represents the maximum number of outliers one believes there might plausibly be in the sample. Define

$$\bar{X}_{i-1} = \sum_{i}^{n} X_j/(n - i + 1)$$

$$S_{i-1} = \sum_{i}^{n} (X_j - \bar{X}_{i-1})^2 + U$$

$$T_{n:i} = |X_i - \bar{X}_{i-1}|/\sqrt{S_{i-1}}$$

Letting $T_{n:i,K,\alpha}$ denote the tabulated fractile, one rejects the null hypothesis of no outliers if

$$\bigcup_{i=1}^{K} (T_{n:i} > T_{n:i,K,\alpha})$$

and concludes that there are k outliers, where k is the largest i such that $T_{n:i} > T_{n:i,K,\alpha}$.

For values outside the range tabulated, there is little error in approximating $T_{n:i,K,\alpha}$ by the α fractile of a B^* statistic (see Appendix 1) based on a sample of size $n - i + 1$ and using ν external degrees of freedom.

The fractiles of $T_{n:i}$ may also be used as approximations when the X_i are not a random sample, but are generated by a linear hypothesis $X_i \sim N(W_i\beta, \sigma^2)$. The external degrees of freedom ν must be reduced by (number of components in $\beta - 1$).

The method used in production of the tables was partly analytic, and partly by simulations of 2000 random samples under each set of conditions studied.

TABLE A4.1 Fractiles of $T_{n:i}$ for maximum of K outliers tested at significance level 0.05

	ν	0					5					15					30				
K	i	n 10	15	20	30	50	10	15	20	30	50	10	15	20	30	50	10	15	20	30	50
2	1	0.783	0.695	0.632	0.547	0.450	0.658	0.611	0.570	0.508	0.430	0.517	0.504	0.486	0.451	0.397	0.411	0.414	0.409	0.392	0.359
2	2	0.789	0.698	0.633	0.546	0.450	0.661	0.610	0.569	0.507	0.430	0.520	0.504	0.485	0.450	0.397	0.417	0.417	0.410	0.393	0.359
3	1	0.804	0.711	0.643	0.554	0.454	0.667	0.620	0.578	0.514	0.433	0.517	0.508	0.490	0.454	0.399	0.408	0.415	0.411	0.394	0.360
3	2	0.806	0.714	0.640	0.552	0.453	0.669	0.620	0.575	0.512	0.433	0.521	0.508	0.489	0.453	0.399	0.415	0.418	0.412	0.395	0.361
3	3	0.796	0.705	0.632	0.546	0.450	0.661	0.612	0.569	0.507	0.430	0.521	0.505	0.486	0.450	0.397	0.417	0.417	0.410	0.393	0.359
4	1	0.825	0.727	0.655	0.561	0.458	0.675	0.629	0.586	0.519	0.436	0.515	0.511	0.494	0.458	0.402	0.403	0.415	0.412	0.396	0.362
4	2	0.832	0.734	0.653	0.559	0.458	0.681	0.630	0.583	0.517	0.436	0.523	0.513	0.493	0.457	0.401	0.412	0.419	0.413	0.397	0.362
4	3	0.820	0.723	0.642	0.554	0.454	0.672	0.621	0.575	0.512	0.432	0.522	0.508	0.489	0.453	0.399	0.416	0.418	0.412	0.394	0.360
4	4	0.811	0.711	0.634	0.546	0.450	0.664	0.613	0.569	0.507	0.430	0.521	0.505	0.486	0.450	0.397	0.417	0.417	0.410	0.393	0.359
5	1	0.844	0.745	0.668	0.568	0.461	0.681	0.638	0.594	0.525	0.440	0.511	0.513	0.497	0.461	0.404	0.395	0.414	0.413	0.398	0.364
5	2	0.854	0.752	0.666	0.567	0.461	0.693	0.642	0.592	0.523	0.439	0.522	0.516	0.497	0.460	0.404	0.407	0.418	0.415	0.398	0.364
5	3	0.837	0.744	0.653	0.559	0.458	0.682	0.633	0.583	0.518	0.436	0.523	0.513	0.493	0.457	0.401	0.412	0.419	0.413	0.397	0.362
5	4	0.839	0.729	0.643	0.552	0.454	0.673	0.621	0.575	0.512	0.433	0.522	0.508	0.489	0.453	0.399	0.415	0.418	0.412	0.395	0.361
5	5	0.841	0.717	0.635	0.546	0.450	0.666	0.613	0.569	0.507	0.430	0.521	0.505	0.486	0.450	0.397	0.417	0.417	0.410	0.393	0.359

TABLE A4.2 Fractiles of $T_{n:i}$ for maximum of K outliers tested at significance level 0.10

		ν	0					5					15					30				
k	i	n	10	15	20	30	50	10	15	20	30	50	10	15	20	30	50	10	15	20	30	50
2	1		0.746	0.658	0.597	0.516	0.426	0.617	0.574	0.536	0.479	0.406	0.480	0.471	0.455	0.424	0.375	0.379	0.385	0.382	0.368	0.338
2	2		0.761	0.669	0.602	0.518	0.428	0.626	0.579	0.539	0.481	0.408	0.489	0.476	0.458	0.426	0.376	0.388	0.390	0.386	0.371	0.340
3	1		0.768	0.673	0.607	0.522	0.429	0.625	0.582	0.543	0.484	0.409	0.479	0.474	0.459	0.427	0.377	0.375	0.385	0.383	0.370	0.340
3	2		0.783	0.688	0.612	0.525	0.431	0.635	0.588	0.546	0.486	0.411	0.490	0.480	0.462	0.429	0.379	0.385	0.391	0.387	0.372	0.342
3	3		0.764	0.673	0.602	0.519	0.427	0.627	0.580	0.539	0.481	0.408	0.487	0.475	0.457	0.425	0.376	0.388	0.390	0.385	0.370	0.340
4	1		0.791	0.689	0.619	0.529	0.432	0.633	0.591	0.550	0.489	0.412	0.476	0.476	0.462	0.430	0.379	0.370	0.385	0.384	0.371	0.341
4	2		0.807	0.699	0.624	0.531	0.435	0.648	0.597	0.554	0.491	0.414	0.491	0.483	0.466	0.432	0.381	0.381	0.390	0.388	0.374	0.343
4	3		0.793	0.692	0.615	0.527	0.430	0.637	0.589	0.546	0.486	0.410	0.489	0.478	0.461	0.429	0.378	0.385	0.390	0.386	0.372	0.341
4	4		0.788	0.683	0.605	0.520	0.428	0.629	0.581	0.539	0.481	0.408	0.488	0.475	0.458	0.426	0.376	0.387	0.390	0.385	0.370	0.340
5	1		0.815	0.706	0.631	0.536	0.436	0.638	0.599	0.558	0.494	0.415	0.471	0.478	0.465	0.433	0.381	0.362	0.384	0.385	0.373	0.343
5	2		0.837	0.725	0.639	0.539	0.437	0.656	0.608	0.562	0.496	0.417	0.488	0.485	0.469	0.436	0.383	0.375	0.390	0.389	0.376	0.345
5	3		0.822	0.704	0.626	0.532	0.434	0.650	0.598	0.554	0.490	0.413	0.489	0.481	0.465	0.431	0.380	0.382	0.390	0.388	0.373	0.343
5	4		0.809	0.698	0.617	0.527	0.430	0.638	0.589	0.546	0.486	0.410	0.489	0.478	0.461	0.429	0.378	0.385	0.390	0.386	0.372	0.341
5	5		0.821	0.692	0.606	0.519	0.427	0.630	0.581	0.539	0.480	0.408	0.488	0.475	0.458	0.426	0.376	0.387	0.390	0.385	0.370	0.340

TABLE A4.3 Fractiles of $T_{n:i}$ for maximum of K outliers tested at significance level 0.01

	ν	0					5					15					30				
K	i	n 10	15	20	30	50	10	15	20	30	50	10	15	20	30	50	10	15	20	30	50
2	1	0.844	0.764	0.700	0.608	0.501	0.731	0.681	0.637	0.568	0.480	0.589	0.571	0.548	0.506	0.444	0.474	0.473	0.464	0.443	0.40
2	2	0.836	0.757	0.691	0.604	0.498	0.726	0.675	0.630	0.563	0.477	0.589	0.568	0.544	0.503	0.441	0.476	0.472	0.463	0.441	0.40
3	1	0.860	0.779	0.711	0.616	0.505	0.741	0.691	0.645	0.574	0.483	0.590	0.575	0.553	0.510	0.446	0.472	0.474	0.466	0.445	0.40
3	2	0.853	0.773	0.702	0.609	0.502	0.734	0.683	0.638	0.569	0.480	0.592	0.573	0.549	0.507	0.444	0.475	0.474	0.465	0.443	0.40
3	3	0.846	0.759	0.693	0.605	0.498	0.729	0.675	0.630	0.564	0.477	0.590	0.568	0.545	0.504	0.441	0.476	0.472	0.463	0.441	0.40
4	1	0.873	0.795	0.724	0.623	0.509	0.749	0.701	0.654	0.580	0.487	0.590	0.579	0.557	0.514	0.449	0.468	0.475	0.468	0.447	0.40
4	2	0.872	0.791	0.716	0.617	0.506	0.746	0.695	0.647	0.575	0.484	0.595	0.578	0.554	0.511	0.447	0.473	0.476	0.467	0.445	0.40
4	3	0.862	0.777	0.704	0.611	0.502	0.738	0.685	0.638	0.569	0.480	0.594	0.574	0.549	0.507	0.444	0.475	0.474	0.465	0.443	0.40
4	4	0.848	0.762	0.694	0.606	0.498	0.728	0.675	0.630	0.564	0.477	0.590	0.568	0.545	0.504	0.441	0.476	0.472	0.463	0.441	0.40
5	1	0.882	0.811	0.736	0.632	0.514	0.755	0.711	0.662	0.586	0.490	0.587	0.583	0.562	0.518	0.452	0.461	0.475	0.470	0.449	0.40
5	2	0.893	0.803	0.728	0.625	0.510	0.758	0.704	0.655	0.581	0.487	0.600	0.582	0.559	0.515	0.449	0.470	0.476	0.469	0.447	0.40
5	3	0.875	0.789	0.716	0.617	0.506	0.748	0.695	0.647	0.575	0.484	0.599	0.579	0.555	0.511	0.447	0.474	0.476	0.467	0.445	0.40
5	4	0.870	0.777	0.705	0.611	0.502	0.739	0.685	0.639	0.569	0.480	0.594	0.574	0.550	0.507	0.444	0.475	0.474	0.465	0.443	0.40
5	5	0.866	0.763	0.695	0.606	0.498	0.729	0.675	0.630	0.564	0.477	0.590	0.568	0.545	0.504	0.441	0.476	0.472	0.463	0.441	0.40

APPENDIX 5

Fractiles for testing for two outliers in normal data

Given $X_1, \ldots, X_n \sim N(\xi, \sigma^2)$ and independently $U \sim \sigma^2 \chi_\nu^2$; let $X_{(1)} \leqslant \ldots \leqslant X_{(n)}$ denote the order statistics of the X_i. Define

$$\bar{X}_{0,i} = \sum_1^{n-i} X_{(i)}/(n-i)$$

$$S_{0,i} = \sum_1^{n-i} (X_{(i)} - \bar{X}_{0,i})^2 + U$$

$$\bar{X} = \sum_1^{n} X_{(i)}/n$$

$$S = \sum_1^{n} (X_{(i)} - \bar{X})^2 + U$$

$$G = S_{0,2}/S \qquad (= L_2 \text{ of Appendix 2})$$

$$M = (X_{(n)} + X_{(n-1)} - 2\bar{X})/\sqrt{S}$$

$$R = (X_{(n-1)} - \bar{X}_{0,1})/\sqrt{S_{0,n-1}}$$

Two outliers on the right are indicated by a small value of G, and large values of M and R. The tabulated fractiles are exact, and are taken from Hawkins (1978a).

For parameters outside the range tabulated, G may be bounded very conservatively by the $\alpha \Big/ \binom{n}{2}$ fractile of a beta distribution with $\frac{1}{2}(n+\nu-3)$ and 1 degrees of freedom. The fractiles of M may be approximated by $T[2(n-2)/\{n(n+\nu-2+T^2)\}]^{1/2}$, where T represents the upper $\binom{n}{2}$ fractile of a t-distribution with $n+\nu-2$ degrees of freedom. The approximation is exact if the resulting fractile exceeds $\{(3n-8)/2n\}^{1/2}$.

The fractiles of R are bounded by those of a B statistic (see Appendix 1) based on a sample of size $n-1$. This bound is very conservative, however.

Critical values of G, M and R

	ν	0				5				15				30			
n	α	0.1	0.05	0.01	0.001	0.1	0.05	0.01	0.001	0.1	0.05	0.01	0.001	0.1	0.05	0.01	0.001
5	G	0.036	0.017	0.002	0.000	0.407	0.330	0.204	0.104	0.696	0.638	0.524	0.397	0.824	0.788	0.710	0.614
	M	1.024	1.050	1.080	1.092	0.785	0.838	0.928	1.003	0.564	0.616	0.712	0.811	0.427	0.471	0.554	0.647
	R	0.810	0.840	0.864	0.866	0.496	0.555	0.661	0.756	0.318	0.361	0.445	0.540	0.232	0.264	0.327	0.402
6	G	0.090	0.055	0.017	0.003	0.423	0.351	0.230	0.127	0.688	0.633	0.525	0.403	0.817	0.781	0.706	0.610
	M	1.044	1.077	1.120	1.144	0.827	0.879	0.966	1.044	0.610	0.661	0.756	0.854	0.468	0.512	0.595	0.688
	R	0.771	0.815	0.869	0.892	0.505	0.560	0.660	0.758	0.336	0.377	0.458	0.550	0.247	0.278	0.340	0.413
7	G	0.146	0.100	0.043	0.012	0.440	0.372	0.255	0.150	0.683	0.632	0.528	0.411	0.810	0.775	0.701	0.610
	M	1.048	1.085	1.138	1.172	0.850	0.901	0.986	1.066	0.641	0.691	0.784	0.881	0.497	0.541	0.623	0.716
	R	0.730	0.780	0.853	0.896	0.506	0.557	0.653	0.751	0.347	0.386	0.463	0.553	0.259	0.288	0.348	0.420
8	G	0.197	0.146	0.074	0.028	0.457	0.393	0.279	0.173	0.682	0.632	0.532	0.419	0.806	0.771	0.699	0.609
	M	1.044	1.083	1.143	1.187	0.863	0.912	0.998	1.078	0.662	0.711	0.802	0.898	0.518	0.562	0.644	0.735
	R	0.695	0.746	0.828	0.888	0.503	0.551	0.643	0.740	0.354	0.392	0.466	0.553	0.267	0.296	0.354	0.424
9	G	0.244	0.189	0.107	0.048	0.474	0.413	0.302	0.195	0.682	0.634	0.538	0.428	0.803	0.768	0.698	0.611
	M	1.036	1.076	1.140	1.193	0.869	0.917	1.002	1.083	0.677	0.725	0.814	0.909	0.535	0.578	0.659	0.749
	R	0.665	0.715	0.801	0.872	0.498	0.544	0.632	0.728	0.359	0.395	0.467	0.551	0.273	0.301	0.358	0.426
10	G	0.284	0.229	0.140	0.070	0.490	0.431	0.324	0.217	0.683	0.637	0.544	0.437	0.799	0.767	0.698	0.613
	M	1.027	1.066	1.134	1.192	0.871	0.918	1.002	1.084	0.688	0.735	0.822	0.916	0.548	0.590	0.670	0.759
	R	0.639	0.687	0.774	0.853	0.493	0.536	0.621	0.715	0.362	0.397	0.466	0.548	0.278	0.305	0.361	0.427
11	G	0.321	0.265	0.172	0.094	0.505	0.449	0.344	0.237	0.686	0.641	0.551	0.447	0.799	0.765	0.698	0.615
	M	1.015	1.055	1.125	1.188	0.870	0.916	1.000	1.082	0.695	0.741	0.827	0.920	0.558	0.599	0.678	0.767
	R	0.616	0.662	0.749	0.832	0.486	0.528	0.609	0.701	0.364	0.397	0.465	0.545	0.282	0.308	0.363	0.428

(*Contd.*)

Critical values of G, M and R (*Contd.*)

ν		0				5				15				30			
n	α	0.1	0.05	0.01	0.001	0.1	0.05	0.01	0.001	0.1	0.05	0.01	0.001	0.1	0.05	0.01	0.001
12	G	0.353	0.298	0.203	0.119	0.520	0.466	0.363	0.257	0.688	0.644	0.557	0.456	0.797	0.765	0.700	0.618
	M	1.002	1.043	1.114	1.181	0.867	0.913	0.995	1.080	0.700	0.746	0.830	0.922	0.566	0.607	0.685	0.772
	R	0.596	0.640	0.725	0.811	0.479	0.519	0.598	0.688	0.364	0.397	0.462	0.540	0.284	0.311	0.364	0.428
13	G	0.383	0.328	0.232	0.143	0.534	0.481	0.382	0.276	0.691	0.649	0.564	0.465	0.797	0.765	0.701	0.621
	M	0.989	1.030	1.102	1.173	0.862	0.908	0.989	1.074	0.703	0.748	0.831	0.922	0.572	0.612	0.689	0.776
	R	0.578	0.621	0.703	0.791	0.473	0.511	0.588	0.675	0.364	0.396	0.459	0.536	0.286	0.312	0.364	0.427
14	G	0.409	0.355	0.259	0.167	0.547	0.496	0.399	0.295	0.695	0.654	0.571	0.474	0.796	0.765	0.702	0.624
	M	0.976	1.018	1.091	1.163	0.857	0.902	0.983	1.068	0.705	0.749	0.831	0.921	0.577	0.617	0.693	0.778
	R	0.562	0.603	0.684	0.771	0.466	0.503	0.577	0.663	0.364	0.395	0.457	0.531	0.289	0.314	0.364	0.426
15	G	0.433	0.380	0.284	0.190	0.559	0.510	0.415	0.312	0.698	0.658	0.578	0.483	0.796	0.766	0.704	0.628
	M	0.964	1.005	1.079	1.153	0.851	0.895	0.976	1.061	0.706	0.749	0.830	0.919	0.581	0.620	0.695	0.780
	R	0.548	0.587	0.665	0.752	0.460	0.496	0.567	0.651	0.363	0.393	0.453	0.526	0.290	0.314	0.364	0.425
20	G	0.525	0.479	0.390	0.292	0.612	0.569	0.485	0.389	0.717	0.682	0.610	0.524	0.799	0.772	0.716	0.645
	M	0.904	0.944	1.019	1.099	0.818	0.860	0.937	1.021	0.700	0.740	0.816	0.901	0.590	0.627	0.698	0.778
	R	0.491	0.524	0.593	0.673	0.431	0.461	0.524	0.600	0.355	0.382	0.436	0.502	0.292	0.314	0.360	0.416
25	G	0.589	0.548	0.467	0.374	0.653	0.615	0.539	0.450	0.735	0.703	0.639	0.560	0.804	0.780	0.728	0.663
	M	0.853	0.893	0.966	1.046	0.785	0.825	0.899	0.981	0.687	0.724	0.797	0.878	0.589	0.624	0.692	0.769
	R	0.450	0.480	0.541	0.614	0.406	0.433	0.490	0.558	0.346	0.370	0.419	0.480	0.290	0.311	0.353	0.406
30	G	0.636	0.600	0.526	0.438	0.685	0.651	0.582	0.499	0.751	0.722	0.664	0.590	0.811	0.788	0.740	0.679
	M	0.810	0.848	0.920	1.000	0.755	0.793	0.864	0.944	0.671	0.707	0.776	0.854	0.585	0.618	0.682	0.757
	R	0.421	0.447	0.502	0.569	0.385	0.410	0.462	0.525	0.335	0.358	0.403	0.460	0.287	0.306	0.346	0.396

APPENDIX 6

Probabilities of the Mosteller test

Given X_{ij}, $j = 1, \ldots, n$; $i = 1, \ldots, m$; let I denote the treatment i containing the maximum order statistic in the combined set of nm observations. The test statistic is the number of observations from treatment I exceeding all observations in all other groups. The probability that this statistic equals r is given by $n!m(nm-r)!/\{(mn)!(n-r)!\}$ which may be computed fairly easily for values outside the range tabulated.

The table is taken from Mosteller (1948). Large values of the test statistic suggest that population I has slipped to the right. An obvious adaptation may be used to test whether any population has slipped to the left.

$m = 2$

	r				
n	2	3	4	5	6
3	0.400	0.100			
5	0.444	0.167	0.048	0.008	
7	0.462	0.192	0.070	0.021	0.005
10	0.474	0.211	0.087	0.033	0.011
15	0.483	0.224	0.100	0.042	0.017
20	0.487	0.231	0.106	0.047	0.020
25	0.490	0.235	0.110	0.050	0.022
∞	0.500	0.250	0.125	0.062	0.031

$m = 3$

	r				
n	2	3	4	5	6
3	0.250	0.036			
5	0.286	0.066	0.011	0.001	
7	0.300	0.079	0.018	0.003	0.0004
10	0.310	0.089	0.023	0.005	0.0011
15	0.318	0.096	0.027	0.007	0.0018
20	0.322	0.100	0.030	0.009	0.0023
25	0.324	0.102	0.031	0.009	0.0026
∞	0.333	0.111	0.037	0.012	0.0041

$m = 4$

	r				
n	2	3	4	5	6
3	0.182	0.018			
5	0.211	0.035	0.004	0.0003	
7	0.222	0.043	0.007	0.0009	0.0001
10	0.231	0.049	0.009	0.0015	0.0002
15	0.237	0.053	0.011	0.0022	0.0004
20	0.241	0.056	0.012	0.0026	0.0005
25	0.242	0.057	0.013	0.0028	0.0006
∞	0.250	0.062	0.016	0.0039	0.0010

$m = 5$

	r			
n	2	3	4	5
3	0.143	0.011		
5	0.167	0.022	0.0020	0.0001
7	0.177	0.027	0.0033	0.0003
10	0.184	0.031	0.0046	0.0006
15	0.189	0.034	0.0056	0.0008
20	0.192	0.035	0.0062	0.0010
25	0.194	0.036	0.0065	0.0011
∞	0.200	0.040	0.0080	0.0016

$m = 6$

	r			
n	2	3	4	5
3	0.118	0.007		
5	0.138	0.015	0.0011	0.0000
7	0.146	0.018	0.0019	0.0001
10	0.152	0.021	0.0026	0.0003
15	0.157	0.023	0.0032	0.0004
20	0.160	0.024	0.0035	0.0005
25	0.161	0.025	0.0037	0.0005
∞	0.167	0.028	0.0046	0.0008

APPENDIX 7

Fractiles of the Doornbos test

Given X_{ij}, $j = 1, \ldots, n; i = 1, \ldots, m$; let R_i be the sum of the ranks of the $\{X_{ij}\}$ in the combined sample of size nm. The test concludes that the treatment corresponding to $R_{(m)}$, the largest R_i, has slipped to the right if $R_{(m)}$ is sufficiently large. The values tabulated were obtained from the Boneferroni inequality.

Outside the range of values tabulated, one may use the approximation

$$\tfrac{1}{2}n(nm + 1) + U_{1 - \alpha/m}\{n^2(m - 1)(mn + 1)/12\}^{1/2}$$

where $U_{1 - \alpha/m}$ is the upper α/m fractile of N(0, 1). Comparison of this approximation with the exact values in the table suggests that the approximation is always slightly conservative, but is most accurate for large values of n and small values of m.

Fractiles of the Doornbos statistic for n observations per group and m groups

n	m	α 0.10	α 0.05	α 0.01	n	m	α 0.10	α 0.05	α 0.01	n	m	α 0.10	α 0.05	α 0.01
3	3	23	24		3	4	31	32		3	5	39	40	
3	6	47	49	51	3	7	55	57	60	3	8	63	65	68
3	9	71	73	77	3	10	80	82	85	3	15	121	124	129
4	3	38	39	42	4	4	51	53	56	4	5	64	67	71
4	6	78	81	86	4	7	92	95	100	4	8	105	109	115
4	9	119	123	130	4	10	133	137	145	4	15	203	209	219
5	3	56	58	62	5	4	76	79	84	5	5	96	99	106
5	6	116	120	128	5	7	137	141	150	5	8	157	162	172
5	9	178	184	194	5	10	199	205	217	5	15	304	313	329
6	3	78	80	86	6	4	105	109	116	6	5	133	138	147

APPENDIX 8

Fractiles of the Wilks statistics

Given $\mathbf{X}_1, \mathbf{X}_2, \ldots, \mathbf{X}_n \sim MN(\xi, \Sigma)$
Re-order the vectors so that $\mathbf{X}_1, \mathbf{X}_2, \ldots, \mathbf{X}_k$ are the suspected outliers.

Define $\bar{\mathbf{X}}_i = \sum_{i+1}^{n} \mathbf{X}_j/(n-i)$

$$\mathbf{A}_i = \sum_{i+1}^{n} (\mathbf{X}_j - \bar{\mathbf{X}}_i)(\mathbf{X}_j - \bar{\mathbf{X}}_i)'$$

The test statistic is $|\mathbf{A}_i|/|\mathbf{A}_0|$ and small values suggest that the suspected observations are indeed contaminants.

The fractiles listed have the property that over all possible partitioning of the observations into two groups of sizes k and $n-k$, the expected number of test statistics exceeding the tabulated value is α per sample. It follows from the Bonferroni inequality that these bounds are conservative (and very much so far $k > 1$), having a size $\leqslant \alpha$.

The tables extend only to $k = 1$ and $k = 2$. For values not covered by the tables, one may use as a conservative approximation the $\alpha \Big/ \binom{n}{k}$ fractile of the Wilks' Λ statistic; however, the common χ^2 approximations to this are likely to yield rather poor estimates of these fractiles in view of the miniscule tail areas involved.

$\alpha = 0.1$ $k = 1$

n	p = 1	2	3	4	5
5	0.12702	0.02000	0.00025		
6	0.20317	0.06525	0.01114	0.00012	
7	0.26960	0.11952	0.04172	0.00717	0.00007
8	0.32610	0.17329	0.08282	0.02959	0.00502
9	0.37418	0.22314	0.12675	0.06216	0.02234
10	0.41540	0.26827	0.16979	0.09888	0.04901
11	0.45106	0.30878	0.21038	0.13629	0.08032
12	0.48221	0.34511	0.24801	0.17267	0.11319
13	0.50966	0.37776	0.28264	0.20723	0.14593
14	0.53405	0.40219	0.31442	0.23968	0.17764
15	0.55586	0.43383	0.34358	0.26995	0.20789
20	0.63789	0.53616	0.45832	0.39255	0.33484
25	0.69224	0.60535	0.53774	0.47967	0.42784
30	0.73119	0.65540	0.59583	0.54420	0.49768
35	0.76063	0.69342	0.64023	0.59385	0.55182
40	0.78375	0.72335	0.67532	0.63326	0.59499
45	0.80245	0.74758	0.70379	0.66532	0.63023
50	0.81792	0.76763	0.72738	0.69195	0.65955
60	0.84208	0.79895	0.76430	0.73369	0.70561
70	0.86017	0.82238	0.79193	0.76497	0.74019
80	0.87426	0.84061	0.81344	0.78935	0.76717
90	0.88559	0.85524	0.83070	0.80891	0.78883
100	0.89490	0.86725	0.84487	0.82497	0.80662

$\alpha = 0.1$ $k = 2$

n	p = 1	2	3	4	5
5	0.10000	0.00501			
6	0.18821	0.04791	0.00223		
7	0.26269	0.10904	0.02872	0.00119	
8	0.32402	0.16955	0.07368	0.01927	0.00072
9	0.37493	0.22444	0.12283	0.05397	0.01386
10	0.41780	0.27305	0.17039	0.09468	0.04161
11	0.45442	0.31592	0.21449	0.13599	0.07599
12	0.48609	0.35381	0.25474	0.17565	0.11219
13	0.51380	0.38747	0.29126	0.21282	0.14791
14	0.53827	0.41753	0.32440	0.24728	0.18212
15	0.56006	0.44453	0.35452	0.27909	0.21439
20	0.64140	0.54676	0.47079	0.40506	0.34649
25	0.69494	0.61487	0.54966	0.49243	0.44064
30	0.73323	0.66380	0.60677	0.55631	0.51026
35	0.76216	0.70082	0.65015	0.60508	0.56375
40	0.78487	0.72990	0.68431	0.64361	0.60614
45	0.80328	0.75343	0.71196	0.67485	0.64061
50	0.81850	0.77288	0.73485	0.70074	0.66921
60	0.84231	0.80327	0.77061	0.74124	0.71401
70	0.86017	0.82600	0.79736	0.77155	0.74758
80	0.87410	0.84370	0.81817	0.79514	0.77372
90	0.88531	0.85791	0.83488	0.81407	0.79470
100	0.89454	0.86960	0.84859	0.82961	0.81193

$\alpha = 0.05, \quad k = 1$

	p				
n	1	2	3	4	5
5	0.08083	0.01000	0.00006		
6	0.14528	0.04110	0.00556	0.00003	
7	0.20661	0.08452	0.02620	0.00358	0.00002
8	0.26161	0.13133	0.05831	0.01856	0.00250
9	0.31006	0.17711	0.09559	0.04367	0.01400
10	0.35261	0.22007	0.13408	0.07438	0.03440
11	0.39008	0.25965	0.17171	0.10731	0.06032
12	0.42324	0.29584	0.20751	0.14050	0.08896
13	0.45277	0.32886	0.24112	0.17285	0.11850
14	0.47921	0.35897	0.27245	0.20383	0.14786
15	0.50302	0.38650	0.30154	0.23319	0.17642
20	0.59365	0.49417	0.41876	0.35558	0.30060
25	0.65443	0.56838	0.50188	0.44513	0.39477
30	0.69824	0.62260	0.56347	0.51248	0.46675
35	0.73146	0.66402	0.61090	0.56478	0.52314
40	0.75758	0.69675	0.64857	0.60654	0.56843
45	0.77872	0.72331	0.67924	0.64067	0.60557
50	0.79621	0.74532	0.70472	0.66909	0.63659
60	0.82354	0.77976	0.74468	0.71377	0.68548
70	0.84397	0.80554	0.77464	0.74735	0.72232
80	0.85989	0.82561	0.79800	0.77356	0.75111
90	0.87266	0.84172	0.81674	0.79462	0.77426
100	0.88315	0.85494	0.83214	0.81192	0.79329

$\alpha = 0.05 \quad k = 2$

	p				
n	1	2	3	4	5
5	0.07071	0.00250			
6	0.14938	0.03371	0.00111		
7	0.22090	0.08601	0.02019	0.00060	
8	0.28207	0.14167	0.05800	0.01355	0.00036
9	0.33402	0.19419	0.10237	0.04245	0.00975
10	0.37841	0.24188	0.14702	0.07881	0.03273
11	0.41670	0.28462	0.18945	0.11716	0.06322
12	0.45006	0.32286	0.22884	0.15489	0.09657
13	0.47939	0.35711	0.26504	0.19085	0.13029
14	0.50539	0.38792	0.29819	0.22462	0.16313
15	0.52863	0.41573	0.32853	0.25609	0.19451
20	0.61578	0.52205	0.44722	0.38281	0.32571
25	0.67339	0.59354	0.52878	0.47215	0.42108
30	0.71465	0.64513	0.58821	0.53799	0.49230
35	0.74583	0.68424	0.63351	0.58850	0.54730
40	0.77032	0.71502	0.66926	0.62850	0.59106
45	0.79013	0.73992	0.69824	0.66101	0.62671
50	0.80652	0.76052	0.72224	0.68798	0.65635
60	0.83213	0.79271	0.75978	0.73021	0.70284
70	0.85132	0.81678	0.78787	0.76185	0.73772
80	0.86627	0.83553	0.80974	0.78650	0.76491
90	0.87829	0.85057	0.82728	0.80627	0.78674
100	0.88818	0.86292	0.84169	0.82251	0.80467

$\alpha = 0.01, \quad k = 1$

	p				
n	1	2	3	4	5
5	0.02795	0.00200	0.00000		
6	0.06592	0.01406	0.00111	0.00000	
7	0.11027	0.03780	0.00893	0.00071	0.00000
8	0.15547	0.06899	0.02593	0.00632	0.00050
9	0.19888	0.10357	0.04987	0.01937	0.00476
10	0.23942	0.13895	0.07781	0.03866	0.01523
11	0.27678	0.17364	0.10755	0.06200	0.03128
12	0.31103	0.20689	0.13765	0.08757	0.05126
13	0.34238	0.23835	0.16726	0.11407	0.07362
14	0.37107	0.26790	0.19590	0.14065	0.09723
15	0.39738	0.29556	0.22330	0.16678	0.12128
20	0.50112	0.40892	0.34019	0.28354	0.23506
25	0.57334	0.49102	0.42815	0.37513	0.32861
30	0.62644	0.55263	0.49547	0.44663	0.40320
35	0.66716	0.60048	0.54835	0.50344	0.46318
40	0.69943	0.63870	0.59091	0.54949	0.51217
45	0.72567	0.66994	0.62588	0.58754	0.55284
50	0.74745	0.69598	0.65514	0.61947	0.58711
60	0.78157	0.73694	0.70133	0.67009	0.64162
70	0.80715	0.76775	0.73620	0.70843	0.68306
80	0.82708	0.79181	0.76348	0.73850	0.71563
90	0.84307	0.81114	0.78544	0.76274	0.74192
100	0.85621	0.82704	0.80352	0.78271	0.76361

$\alpha = 0.01, \quad k = 2$

	p				
n	1	2	3	4	5
5	0.03162	0.00050			
6	0.08736	0.01498	0.00022		
7	0.14772	0.04982	0.00896	0.00012	
8	0.20444	0.09374	0.03349	0.00601	0.00007
9	0.25544	0.13926	0.06744	0.02449	0.00433
10	0.30069	0.18308	0.10490	0.05181	0.01887
11	0.34076	0.22397	0.14265	0.08337	0.04152
12	0.37636	0.26160	0.17912	0.11626	0.06861
13	0.40812	0.29606	0.21363	0.14889	0.09761
14	0.43660	0.32757	0.24595	0.18044	0.12699
15	0.46228	0.35641	0.27605	0.21050	0.15589
20	0.56016	0.46935	0.39764	0.33655	0.28296
25	0.62588	0.54722	0.48390	0.42895	0.37975
30	0.67329	0.60409	0.54778	0.49839	0.45370
35	0.70925	0.64753	0.59694	0.55228	0.51159
40	0.73753	0.68184	0.63596	0.59527	0.55804
45	0.76042	0.70968	0.66772	0.63039	0.59611
50	0.77937	0.73276	0.69411	0.65962	0.62789
60	0.80897	0.76887	0.73547	0.70556	0.67795
70	0.83111	0.79590	0.76649	0.74009	0.71566
80	0.84835	0.81695	0.79066	0.76702	0.74512
90	0.86219	0.83384	0.81007	0.78866	0.76880
100	0.87356	0.84771	0.82601	0.80645	0.78828

APPENDIX 9

Fractiles of $X_{(n)}/W$ for samples from the chi-squared distribution

Given $X_0 \sim \sigma^2\chi^2_m$, $X_1, \ldots X_n \sim \sigma^2\chi^2_\nu$ all mutually independent; $X_{(n)} = \max(X_1, \ldots, X_n)$, $W = \sum_0^n X_i$, the statistic $X_{(n)}/W$ is used to test for slippage on the right. Its exact distribution is known, but rather tedious to compute. The fractiles listed here are slightly conservative approximations obtained from Bonferroni's inequality:

$$\Pr[X_{(n)}/W > k] \leqslant n \int_k^1 g(s)\,ds,$$

where

$$g(s) = \frac{\Gamma[\frac{1}{2}(\nu n + m)]s^{1/2\nu - 1}(1 - s)^{1/2\{\nu(n-1)+m\} - 1}}{\Gamma(\frac{1}{2}\nu)\Gamma[\frac{1}{2}\{\nu(n-1) + m\}]}$$

Choosing k as the α/n fractile of the beta distribution, $g(s)$ then yields

$$\alpha - \tfrac{1}{2}\alpha^2 < \Pr[X_{(n)}/W > k] \leqslant \alpha$$

with equality on the right side if and only if $k > 0.5$.

$\nu = 1$

n \ m	0				10				20				30				50			
α	0.100	0.050	0.010	0.001	0.100	0.050	0.010	0.001	0.100	0.050	0.010	0.001	0.100	0.050	0.010	0.001	0.100	0.050	0.010	0.001
3	0.9344	0.9669	0.9933	0.9993	0.3249	0.3917	0.5260	0.6717	0.1898	0.2338	0.3299	0.4500	0.1339	0.1663	0.2392	0.3353	0.0842	0.1053	0.1540	0.2213
4	0.8533	0.9065	0.9676	0.9930	0.3304	0.3922	0.5176	0.6561	0.2000	0.2420	0.3336	0.4486	0.1432	0.1746	0.2451	0.3378	0.0912	0.1120	0.1598	0.2255
5	0.7783	0.8413	0.9279	0.9770	0.3298	0.3876	0.5061	0.6395	0.2056	0.2458	0.3337	0.4448	0.1491	0.1796	0.2479	0.3378	0.0962	0.1166	0.1634	0.2278
6	0.7141	0.7807	0.8828	0.9530	0.3260	0.3807	0.4934	0.6229	0.2085	0.2472	0.3318	0.4389	0.1530	0.1826	0.2490	0.3367	0.0998	0.1199	0.1658	0.2290
7	0.6599	0.7270	0.8376	0.9239	0.3207	0.3725	0.4803	0.6054	0.2096	0.2469	0.3286	0.4325	0.1555	0.1844	0.2490	0.3347	0.1025	0.1222	0.1674	0.2295
8	0.6138	0.6798	0.7945	0.8922	0.3145	0.3639	0.4674	0.5893	0.2097	0.2457	0.3248	0.4258	0.1571	0.1852	0.2481	0.3320	0.1045	0.1239	0.1683	0.2295
9	0.5742	0.6385	0.7544	0.8602	0.3079	0.3552	0.4547	0.5732	0.2089	0.2438	0.3205	0.4188	0.1579	0.1853	0.2468	0.3284	0.1061	0.1251	0.1688	0.2285
10	0.5399	0.6020	0.7175	0.8284	0.3011	0.3465	0.4425	0.5580	0.2076	0.2414	0.3159	0.4120	0.1583	0.1850	0.2451	0.3252	0.1072	0.1260	0.1689	0.2279
11	0.5099	0.5697	0.6837	0.7979	0.2943	0.3380	0.4308	0.5432	0.2059	0.2388	0.3112	0.4047	0.1582	0.1844	0.2431	0.3213	0.1081	0.1266	0.1688	0.2269
12	0.4834	0.5410	0.6528	0.7688	0.2877	0.3297	0.4195	0.5294	0.2040	0.2359	0.3064	0.3978	0.1579	0.1834	0.2409	0.3176	0.1087	0.1269	0.1685	0.2256
13	0.4598	0.5152	0.6245	0.7412	0.2812	0.3217	0.4088	0.5161	0.2018	0.2329	0.3016	0.3910	0.1573	0.1823	0.2386	0.3137	0.1091	0.1270	0.1680	0.2246
14	0.4386	0.4919	0.5985	0.7148	0.2748	0.3141	0.3985	0.5034	0.1995	0.2298	0.2968	0.3843	0.1565	0.1810	0.2361	0.3101	0.1093	0.1270	0.1674	0.2230
15	0.4195	0.4709	0.5747	0.6905	0.2687	0.3067	0.3887	0.4913	0.1972	0.2267	0.2921	0.3777	0.1556	0.1796	0.2337	0.3061	0.1094	0.1269	0.1667	0.2215
20	0.3465	0.3894	0.4799	0.5881	0.2414	0.2742	0.3461	0.4382	0.1850	0.2113	0.2699	0.3476	0.1499	0.1717	0.2210	0.2875	0.1086	0.1249	0.1621	0.2134
25	0.2969	0.3337	0.4130	0.5116	0.2190	0.2479	0.3119	0.3954	0.1734	0.1971	0.2502	0.3214	0.1434	0.1634	0.2088	0.2704	0.1066	0.1218	0.1567	0.2051
30	0.2607	0.2929	0.3632	0.4531	0.2004	0.2264	0.2841	0.3604	0.1627	0.1843	0.2331	0.2987	0.1369	0.1554	0.1975	0.2549	0.1039	0.1183	0.1512	0.1967
35	0.2331	0.2617	0.3247	0.4069	0.1849	0.2084	0.2610	0.3313	0.1532	0.1731	0.2180	0.2793	0.1307	0.1480	0.1872	0.2410	0.1011	0.1146	0.1458	0.1892
40	0.2113	0.2369	0.2940	0.3691	0.1717	0.1932	0.2415	0.3066	0.1446	0.1631	0.2048	0.2617	0.1249	0.1410	0.1778	0.2285	0.0981	0.1110	0.1406	0.1819
45	0.1935	0.2168	0.2690	0.3386	0.1604	0.1802	0.2249	0.2854	0.1370	0.1542	0.1932	0.2466	0.1195	0.1347	0.1693	0.2171	0.0952	0.1075	0.1357	0.1750
50	0.1787	0.2000	0.2481	0.3125	0.1506	0.1689	0.2105	0.2672	0.1301	0.1462	0.1828	0.2331	0.1145	0.1288	0.1615	0.2068	0.0924	0.1041	0.1310	0.1686
60	0.1554	0.1737	0.2151	0.2715	0.1344	0.1504	0.1869	0.2369	0.1183	0.1326	0.1652	0.2104	0.1057	0.1185	0.1480	0.1889	0.0871	0.0978	0.1225	0.1570
70	0.1379	0.1539	0.1903	0.2402	0.1215	0.1357	0.1682	0.2133	0.1085	0.1214	0.1508	0.1916	0.0981	0.1098	0.1366	0.1740	0.0822	0.0921	0.1149	0.1470
80	0.1242	0.1385	0.1709	0.2158	0.1110	0.1238	0.1532	0.1939	0.1003	0.1120	0.1388	0.1761	0.0915	0.1022	0.1268	0.1613	0.0778	0.0870	0.1082	0.1379
90	0.1132	0.1260	0.1553	0.1960	0.1023	0.1140	0.1407	0.1781	0.0933	0.1040	0.1286	0.1631	0.0858	0.0957	0.1184	0.1504	0.0739	0.0825	0.1022	0.1301
100	0.1041	0.1157	0.1424	0.1797	0.0950	0.1057	0.1302	0.1646	0.0873	0.0972	0.1199	0.1519	0.0808	0.0900	0.1111	0.1409	0.0703	0.0783	0.0969	0.1230

$\nu = 2$

m	0				10				20				30				50			
n \ α	0.100	0.050	0.010	0.001	0.100	0.050	0.010	0.001	0.100	0.050	0.010	0.001	0.100	0.050	0.010	0.001	0.100	0.050	0.010	0.001
3	0.8174	0.8709	0.9423	0.9817	0.3848	0.4428	0.5573	0.6814	0.2468	0.2891	0.3783	0.4871	0.1813	0.2140	0.2850	0.3753	0.1184	0.1407	0.1904	0.2568
4	0.7076	0.7679	0.8643	0.9371	0.3694	0.4218	0.5271	0.6454	0.2471	0.2861	0.3693	0.4716	0.1853	0.2161	0.2831	0.3693	0.1234	0.1449	0.1926	0.2564
5	0.6239	0.6838	0.7885	0.8810	0.3525	0.4005	0.4987	0.6121	0.2438	0.2803	0.3585	0.4557	0.1861	0.2152	0.2790	0.3613	0.1262	0.1468	0.1929	0.2545
6	0.5591	0.6161	0.7218	0.8245	0.3360	0.3804	0.4725	0.5810	0.2389	0.2732	0.3472	0.4401	0.1851	0.2129	0.2737	0.3527	0.1276	0.1475	0.1920	0.2517
7	0.5074	0.5612	0.6644	0.7714	0.3204	0.3619	0.4487	0.5527	0.2332	0.2657	0.3360	0.4250	0.1832	0.2097	0.2680	0.3437	0.1281	0.1473	0.1905	0.2485
8	0.4653	0.5157	0.6152	0.7230	0.3059	0.3449	0.4271	0.5271	0.2272	0.2581	0.3251	0.4106	0.1806	0.2060	0.2620	0.3354	0.1280	0.1467	0.1885	0.2451
9	0.4302	0.4775	0.5727	0.6796	0.2926	0.3293	0.4074	0.5036	0.2212	0.2506	0.3147	0.3970	0.1777	0.2021	0.2560	0.3271	0.1275	0.1456	0.1863	0.2412
10	0.4005	0.4450	0.5358	0.6406	0.2803	0.3151	0.3895	0.4824	0.2152	0.2434	0.3048	0.3842	0.1746	0.1981	0.2501	0.3187	0.1267	0.1443	0.1839	0.2372
11	0.3750	0.4169	0.5036	0.6054	0.2690	0.3020	0.3730	0.4622	0.2094	0.2364	0.2954	0.3719	0.1714	0.1941	0.2443	0.3110	0.1257	0.1428	0.1814	0.2334
12	0.3529	0.3924	0.4751	0.5742	0.2586	0.2900	0.3580	0.4443	0.2039	0.2297	0.2865	0.3606	0.1682	0.1901	0.2387	0.3032	0.1245	0.1412	0.1788	0.2295
13	0.3334	0.3709	0.4498	0.5459	0.2490	0.2790	0.3441	0.4269	0.1985	0.2233	0.2781	0.3496	0.1650	0.1861	0.2332	0.2959	0.1233	0.1395	0.1762	0.2261
14	0.3162	0.3517	0.4272	0.5201	0.2401	0.2688	0.3313	0.4115	0.1933	0.2173	0.2702	0.3398	0.1618	0.1823	0.2280	0.2890	0.1219	0.1378	0.1736	0.2222
15	0.3009	0.3346	0.4069	0.4968	0.2318	0.2593	0.3195	0.3973	0.1884	0.2115	0.2627	0.3302	0.1587	0.1785	0.2229	0.2822	0.1206	0.1361	0.1710	0.2185
20	0.2434	0.2705	0.3297	0.4062	0.1981	0.2209	0.2715	0.3379	0.1670	0.1867	0.2306	0.2892	0.1443	0.1616	0.2003	0.2527	0.1134	0.1273	0.1586	0.2017
25	0.2055	0.2281	0.2782	0.3443	0.1734	0.1929	0.2365	0.2949	0.1499	0.1671	0.2056	0.2576	0.1320	0.1473	0.1818	0.2285	0.1066	0.1191	0.1476	0.1868
30	0.1785	0.1979	0.2412	0.2992	0.1544	0.1715	0.2098	0.2617	0.1361	0.1513	0.1856	0.2324	0.1216	0.1353	0.1664	0.2090	0.1002	0.1117	0.1378	0.1738
35	0.1583	0.1753	0.2134	0.2649	0.1395	0.1546	0.1888	0.2353	0.1247	0.1383	0.1693	0.2118	0.1127	0.1251	0.1534	0.1924	0.0945	0.1051	0.1292	0.1626
40	0.1424	0.1575	0.1916	0.2378	0.1273	0.1409	0.1718	0.2139	0.1151	0.1275	0.1557	0.1943	0.1050	0.1164	0.1424	0.1783	0.0894	0.0992	0.1216	0.1526
45	0.1296	0.1432	0.1740	0.2163	0.1172	0.1296	0.1577	0.1963	0.1070	0.1184	0.1443	0.1800	0.0984	0.1089	0.1329	0.1660	0.0847	0.0939	0.1148	0.1438
50	0.1191	0.1315	0.1596	0.1982	0.1087	0.1201	0.1459	0.1816	0.1000	0.1105	0.1344	0.1675	0.0925	0.1023	0.1246	0.1555	0.0806	0.0891	0.1087	0.1360
60	0.1028	0.1132	0.1371	0.1701	0.0951	0.1049	0.1271	0.1579	0.0885	0.0977	0.1185	0.1475	0.0828	0.0914	0.1109	0.1382	0.0733	0.0809	0.0984	0.1228
70	0.0906	0.0997	0.1204	0.1494	0.0847	0.0933	0.1128	0.1399	0.0796	0.0876	0.1060	0.1317	0.0750	0.0826	0.1000	0.1244	0.0673	0.0742	0.0899	0.1119
80	0.0811	0.0892	0.1075	0.1332	0.0765	0.0841	0.1015	0.1258	0.0724	0.0796	0.0960	0.1191	0.0686	0.0755	0.0912	0.1132	0.0623	0.0685	0.0828	0.1029
90	0.0736	0.0808	0.0972	0.1204	0.0698	0.0766	0.0923	0.1143	0.0664	0.0729	0.0879	0.1089	0.0633	0.0695	0.0838	0.1039	0.0579	0.0636	0.0768	0.0952
100	0.0674	0.0739	0.0888	0.1098	0.0643	0.0705	0.0848	0.1047	0.0614	0.0674	0.0810	0.1002	0.0588	0.0645	0.0776	0.0961	0.0542	0.0595	0.0716	0.0886

$v = 3$

m	0				10				20				30				50			
n \ α	0.100	0.050	0.0100	0.001	0.100	0.050	0.010	0.001	0.100	0.050	0.010	0.001	0.100	0.050	0.010	0.001	0.100	0.050	0.010	0.001
3	0.7430	0.7977	0.8831	0.9462	0.4110	0.4630	0.5653	0.6771	0.2804	0.3206	0.4042	0.5045	0.2125	0.2446	0.3132	0.3992	0.1430	0.1657	0.2155	0.2802
4	0.6286	0.6839	0.7814	0.8702	0.3813	0.4276	0.5209	0.6268	0.2718	0.3082	0.3848	0.4786	0.2108	0.2405	0.3042	0.3847	0.1454	0.1669	0.2139	0.2757
5	0.5462	0.5981	0.6957	0.7944	0.3545	0.3964	0.4824	0.5832	0.2613	0.2948	0.3659	0.4541	0.2067	0.2344	0.2941	0.3703	0.1457	0.1660	0.2108	0.2636
6	0.4842	0.5321	0.6258	0.7270	0.3309	0.3692	0.4491	0.5448	0.2506	0.2817	0.3482	0.4314	0.2015	0.2275	0.2838	0.3564	0.1447	0.1641	0.2069	0.2636
7	0.4359	0.4800	0.5685	0.6685	0.3101	0.3455	0.4200	0.5108	0.2402	0.2693	0.3318	0.4109	0.1959	0.2204	0.2738	0.3427	0.1430	0.1616	0.2026	0.2573
8	0.3970	0.4377	0.5209	0.6180	0.2918	0.3247	0.3946	0.4811	0.2304	0.2576	0.3167	0.3921	0.1902	0.2134	0.2642	0.3305	0.1409	0.1588	0.1982	0.2508
9	0.3651	0.4027	0.4810	0.5742	0.2756	0.3064	0.3721	0.4545	0.2211	0.2469	0.3028	0.3750	0.1845	0.2066	0.2550	0.3183	0.1386	0.1558	0.1938	0.2445
10	0.3383	0.3733	0.4469	0.5363	0.2612	0.2900	0.3521	0.4307	0.2125	0.2369	0.2901	0.3589	0.1791	0.2001	0.2465	0.3076	0.1362	0.1527	0.1893	0.2383
11	0.3155	0.3482	0.4175	0.5031	0.2482	0.2754	0.3342	0.4093	0.2045	0.2277	0.2783	0.3444	0.1738	0.1940	0.2383	0.2968	0.1337	0.1496	0.1850	0.2324
12	0.2958	0.3264	0.3919	0.4737	0.2366	0.2623	0.3182	0.3901	0.1971	0.2191	0.2675	0.3310	0.1688	0.1881	0.2307	0.2871	0.1312	0.1465	0.1808	0.2270
13	0.2786	0.3074	0.3695	0.4478	0.2260	0.2504	0.3036	0.3722	0.1901	0.2112	0.2576	0.3183	0.1640	0.1825	0.2235	0.2780	0.1287	0.1435	0.1767	0.2216
14	0.2634	0.2907	0.3495	0.4245	0.2165	0.2397	0.2904	0.3564	0.1837	0.2038	0.2483	0.3070	0.1595	0.1773	0.2168	0.2695	0.1262	0.1406	0.1728	0.2163
15	0.2500	0.2758	0.3318	0.4038	0.2077	0.2298	0.2784	0.3418	0.1776	0.1970	0.2397	0.2964	0.1551	0.1723	0.2104	0.2612	0.1238	0.1377	0.1690	0.2114
20	0.2001	0.2205	0.2654	0.3246	0.1732	0.1912	0.2310	0.2842	0.1527	0.1687	0.2045	0.2525	0.1365	0.1510	0.1834	0.2273	0.1126	0.1247	0.1521	0.1893
25	0.1678	0.1846	0.2220	0.2720	0.1491	0.1642	0.1980	0.2435	0.1341	0.1478	0.1786	0.2203	0.1218	0.1344	0.1627	0.2012	0.1030	0.1138	0.1381	0.1714
30	0.1450	0.1593	0.1914	0.2347	0.1312	0.1442	0.1736	0.2134	0.1197	0.1317	0.1588	0.1956	0.1101	0.1212	0.1463	0.1807	0.0949	0.1045	0.1264	0.1564
35	0.1280	0.1404	0.1685	0.2066	0.1173	0.1288	0.1547	0.1901	0.1083	0.1189	0.1430	0.1761	0.1005	0.1105	0.1330	0.1638	0.0879	0.0967	0.1166	0.1440
40	0.1147	0.1258	0.1507	0.1847	0.1062	0.1165	0.1397	0.1716	0.0989	0.1085	0.1303	0.1602	0.0925	0.1015	0.1220	0.1501	0.0819	0.0900	0.1082	0.1335
45	0.1041	0.1140	0.1364	0.1671	0.0972	0.1065	0.1275	0.1564	0.0911	0.0999	0.1197	0.1470	0.0858	0.0940	0.1127	0.1387	0.0767	0.0842	0.1010	0.1244
50	0.0954	0.1044	0.1248	0.1528	0.0897	0.0981	0.1173	0.1439	0.0845	0.0925	0.1107	0.1360	0.0800	0.0876	0.1048	0.1287	0.0722	0.0791	0.0947	0.1165
60	0.0820	0.0895	0.1068	0.1306	0.0778	0.0850	0.1014	0.1240	0.0740	0.0809	0.0965	0.1182	0.0706	0.0771	0.0921	0.1128	0.0646	0.0706	0.0843	0.1035
70	0.0720	0.0786	0.0935	0.1143	0.0688	0.0751	0.0894	0.1093	0.0659	0.0719	0.0856	0.1047	0.0632	0.0690	0.0822	0.1006	0.0585	0.0638	0.0760	0.0931
80	0.0643	0.0701	0.0832	0.1016	0.0618	0.0674	0.0800	0.0977	0.0595	0.0648	0.0770	0.0941	0.0573	0.0625	0.0743	0.0907	0.0535	0.0583	0.0693	0.0847
90	0.0582	0.0634	0.0751	0.0916	0.0562	0.0611	0.0725	0.0884	0.0543	0.0591	0.0701	0.0854	0.0525	0.0572	0.0678	0.0828	0.0493	0.0537	0.0637	0.0778
100	0.0532	0.0579	0.0685	0.0833	0.0515	0.0560	0.0663	0.0808	0.0499	0.0543	0.0643	0.0783	0.0485	0.0527	0.0624	0.0760	0.0457	0.0497	0.0589	0.0718

$\nu = 4$

n \ m, α	0: 0.100	0: 0.050	0: 0.010	0: 0.001	10: 0.100	10: 0.050	10: 0.010	10: 0.001	20: 0.100	20: 0.050	20: 0.010	20: 0.001	30: 0.100	30: 0.050	30: 0.010	30: 0.001	50: 0.100	50: 0.050	50: 0.010	50: 0.001
3	0.6934	0.7457	0.8335	0.9079	0.4246	0.4723	0.5659	0.6692	0.3032	0.3415	0.4203	0.5145	0.2355	0.2669	0.3331	0.4151	0.1625	0.1854	0.2347	0.2983
4	0.5787	0.6287	0.7212	0.8132	0.3848	0.4268	0.5116	0.6092	0.2869	0.3211	0.3926	0.4797	0.2284	0.2570	0.3177	0.3941	0.1622	0.1834	0.2296	0.2894
5	0.4983	0.5440	0.6329	0.7288	0.3515	0.3891	0.4666	0.5585	0.2707	0.3018	0.3675	0.4491	0.2200	0.2462	0.3026	0.3739	0.1599	0.1798	0.2233	0.2802
6	0.4389	0.4803	0.5635	0.6578	0.3234	0.3576	0.4289	0.5152	0.2556	0.2842	0.3451	0.4213	0.2112	0.2356	0.2883	0.3560	0.1567	0.1755	0.2167	0.2707
7	0.3931	0.4307	0.5080	0.5986	0.2997	0.3310	0.3970	0.4784	0.2419	0.2683	0.3251	0.3973	0.2027	0.2255	0.2750	0.3388	0.1530	0.1709	0.2100	0.2617
8	0.3565	0.3910	0.4627	0.5488	0.2793	0.3082	0.3696	0.4464	0.2294	0.2541	0.3072	0.3753	0.1946	0.2160	0.2627	0.3232	0.1492	0.1662	0.2035	0.2529
9	0.3267	0.3584	0.4251	0.5068	0.2617	0.2885	0.3459	0.4186	0.2181	0.2412	0.2913	0.3558	0.1870	0.2072	0.2514	0.3091	0.1454	0.1615	0.1973	0.2447
10	0.3018	0.3311	0.3934	0.4709	0.2462	0.2713	0.3252	0.3940	0.2079	0.2296	0.2769	0.3382	0.1798	0.1990	0.2410	0.2961	0.1416	0.1570	0.1913	0.2369
11	0.2807	0.3080	0.3663	0.4398	0.2326	0.2561	0.3069	0.3723	0.1986	0.2191	0.2639	0.3222	0.1732	0.1914	0.2314	0.2842	0.1379	0.1527	0.1856	0.2295
12	0.2626	0.2880	0.3428	0.4126	0.2205	0.2426	0.2906	0.3528	0.1900	0.2095	0.2521	0.3081	0.1669	0.1843	0.2215	0.2729	0.1343	0.1485	0.1802	0.2226
13	0.2468	0.2707	0.3223	0.3886	0.2097	0.2305	0.2760	0.3354	0.1823	0.2007	0.2413	0.2949	0.1612	0.1777	0.2143	0.2627	0.1308	0.1446	0.1751	0.2158
14	0.2330	0.2554	0.3043	0.3672	0.1999	0.2197	0.2630	0.3198	0.1751	0.1927	0.2315	0.2827	0.1557	0.1716	0.2067	0.2533	0.1275	0.1407	0.1702	0.2098
15	0.2207	0.2419	0.2882	0.3484	0.1911	0.2099	0.2511	0.3053	0.1685	0.1853	0.2224	0.2715	0.1507	0.1659	0.1996	0.2446	0.1243	0.1371	0.1656	0.2039
20	0.1756	0.1921	0.2288	0.2773	0.1570	0.1721	0.2054	0.2500	0.1421	0.1558	0.1863	0.2274	0.1297	0.1423	0.1765	0.2085	0.1104	0.1213	0.1457	0.1787
25	0.1465	0.1601	0.1904	0.2312	0.1338	0.1463	0.1743	0.2121	0.1231	0.1347	0.1607	0.1958	0.1140	0.1248	0.1460	0.1819	0.0993	0.1088	0.1301	0.1592
30	0.1262	0.1377	0.1635	0.1985	0.1168	0.1276	0.1517	0.1846	0.1088	0.1188	0.1414	0.1723	0.1018	0.1112	0.1345	0.1614	0.0902	0.0986	0.1176	0.1435
35	0.1110	0.1210	0.1435	0.1742	0.1039	0.1133	0.1345	0.1633	0.0976	0.1065	0.1265	0.1538	0.0920	0.1004	0.1194	0.1453	0.0826	0.0902	0.1073	0.1308
40	0.0993	0.1082	0.1281	0.1553	0.0937	0.1020	0.1209	0.1468	0.0886	0.0966	0.1145	0.1392	0.0841	0.0916	0.1087	0.1322	0.0763	0.0832	0.0987	0.1201
45	0.0900	0.0979	0.1158	0.1404	0.0854	0.0929	0.1099	0.1333	0.0812	0.0884	0.1046	0.1270	0.0775	0.0843	0.0999	0.1213	0.0709	0.0772	0.0915	0.1112
50	0.0823	0.0895	0.1057	0.1280	0.0785	0.0854	0.1009	0.1223	0.0750	0.0816	0.0964	0.1169	0.0718	0.0781	0.0924	0.1121	0.0662	0.0720	0.0852	0.1035
60	0.0705	0.0765	0.0902	0.1091	0.0678	0.0736	0.0867	0.1050	0.0652	0.0708	0.0835	0.1011	0.0628	0.0682	0.0805	0.0974	0.0586	0.0636	0.0751	0.0910
70	0.0618	0.0670	0.0788	0.0952	0.0597	0.0647	0.0762	0.0920	0.0578	0.0626	0.0737	0.0891	0.0559	0.0607	0.0704	0.0863	0.0526	0.0571	0.0672	0.0813
80	0.0551	0.0597	0.0701	0.0845	0.0535	0.0579	0.0680	0.0820	0.0519	0.0562	0.0661	0.0797	0.0505	0.0547	0.0642	0.0775	0.0478	0.0518	0.0608	0.0735
90	0.0498	0.0539	0.0631	0.0760	0.0485	0.0524	0.0615	0.0741	0.0472	0.0511	0.0599	0.0721	0.0460	0.0498	0.0584	0.0704	0.0438	0.0474	0.0556	0.0670
100	0.0455	0.0491	0.0575	0.0692	0.0444	0.0479	0.0561	0.0676	0.0433	0.0468	0.0548	0.0660	0.0423	0.0458	0.0536	0.0645	0.0405	0.0438	0.0512	0.0617

$\nu = 5$

n \ m, α	0: 0.100	0: 0.050	0: 0.010	0: 0.001	10: 0.100	10: 0.050	10: 0.010	10: 0.001	20: 0.100	20: 0.050	20: 0.010	20: 0.001	30: 0.100	30: 0.050	30: 0.010	30: 0.001	50: 0.100	50: 0.050	50: 0.010	50: 0.001
3	0.6578	0.7070	0.7933	0.8728	0.4323	0.4766	0.5636	0.6603	0.3198	0.3563	0.4310	0.5197	0.2534	0.2839	0.3479	0.4266	0.1788	0.2016	0.2504	0.3124
4	0.5438	0.5894	0.6761	0.7667	0.3849	0.4236	0.5019	0.5926	0.2969	0.3292	0.3964	0.4783	0.2414	0.2689	0.3270	0.3996	0.1756	0.1965	0.2417	0.2999
5	0.4654	0.5063	0.5875	0.6786	0.3470	0.3814	0.4524	0.5372	0.2761	0.3052	0.3665	0.4424	0.2291	0.2542	0.3076	0.3749	0.1708	0.1903	0.2325	0.2871
6	0.4079	0.4447	0.5195	0.6068	0.3161	0.3471	0.4120	0.4916	0.2577	0.2842	0.3406	0.4114	0.2174	0.2405	0.2900	0.3535	0.1655	0.1838	0.2234	0.2751
7	0.3640	0.3972	0.4659	0.5487	0.2904	0.3187	0.3784	0.4526	0.2414	0.2658	0.3181	0.3843	0.2065	0.2279	0.2702	0.3339	0.1601	0.1773	0.2147	0.2636
8	0.3292	0.3594	0.4227	0.5002	0.2689	0.2948	0.3500	0.4195	0.2271	0.2497	0.2984	0.3606	0.1965	0.2165	0.2509	0.3161	0.1548	0.1710	0.2065	0.2532
9	0.3009	0.3285	0.3870	0.4599	0.2504	0.2744	0.3257	0.3911	0.2144	0.2355	0.2810	0.3398	0.1874	0.2062	0.2470	0.3003	0.1497	0.1650	0.1988	0.2434
10	0.2774	0.3028	0.3572	0.4257	0.2345	0.2568	0.3048	0.3662	0.2031	0.2228	0.2656	0.3212	0.1791	0.1967	0.2353	0.2861	0.1448	0.1594	0.1916	0.2344
11	0.2575	0.2811	0.3318	0.3964	0.2206	0.2414	0.2864	0.3445	0.1929	0.2115	0.2519	0.3047	0.1714	0.1881	0.2207	0.2729	0.1401	0.1541	0.1848	0.2257
12	0.2405	0.2624	0.3099	0.3711	0.2083	0.2278	0.2703	0.3254	0.1838	0.2013	0.2395	0.2897	0.1644	0.1802	0.2150	0.2612	0.1357	0.1490	0.1785	0.2178
13	0.2257	0.2463	0.2909	0.3486	0.1975	0.2158	0.2559	0.3081	0.1755	0.1920	0.2284	0.2763	0.1579	0.1730	0.2002	0.2502	0.1315	0.1443	0.1726	0.2104
14	0.2128	0.2321	0.2741	0.3289	0.1877	0.2051	0.2430	0.2929	0.1679	0.1837	0.2183	0.2641	0.1519	0.1663	0.1980	0.2402	0.1276	0.1398	0.1671	0.2035
15	0.2013	0.2195	0.2593	0.3115	0.1790	0.1954	0.2315	0.2790	0.1611	0.1760	0.2090	0.2528	0.1464	0.1601	0.1905	0.2310	0.1239	0.1357	0.1619	0.1969
20	0.1594	0.1735	0.2048	0.2465	0.1456	0.1586	0.1875	0.2261	0.1339	0.1460	0.1729	0.2090	0.1240	0.1353	0.1604	0.1941	0.1081	0.1180	0.1401	0.1699
25	0.1326	0.1441	0.1699	0.2045	0.1232	0.1339	0.1580	0.1904	0.1150	0.1251	0.1477	0.1784	0.1078	0.1173	0.1307	0.1677	0.0959	0.1044	0.1236	0.1496
30	0.1139	0.1236	0.1455	0.1751	0.1070	0.1162	0.1369	0.1649	0.1009	0.1096	0.1292	0.1558	0.0955	0.1038	0.1223	0.1477	0.0862	0.0937	0.1106	0.1337
35	0.1000	0.1085	0.1274	0.1533	0.0948	0.1028	0.1209	0.1455	0.0901	0.0977	0.1149	0.1384	0.0858	0.0931	0.1096	0.1321	0.0784	0.0851	0.1002	0.1208
40	0.0893	0.0968	0.1136	0.1365	0.0852	0.0923	0.1084	0.1303	0.0814	0.0882	0.1036	0.1248	0.0780	0.0845	0.0903	0.1196	0.0719	0.0779	0.0916	0.1103
45	0.0808	0.0875	0.1025	0.1230	0.0774	0.0838	0.0983	0.1182	0.0744	0.0805	0.0944	0.1136	0.0715	0.0774	0.0908	0.1093	0.0664	0.0720	0.0844	0.1016
50	0.0738	0.0799	0.0935	0.1122	0.0711	0.0769	0.0900	0.1080	0.0685	0.0741	0.0868	0.1042	0.0661	0.0715	0.0808	0.1006	0.0618	0.0669	0.0783	0.0942
60	0.0631	0.0682	0.0796	0.0955	0.0611	0.0660	0.0771	0.0924	0.0593	0.0640	0.0748	0.0896	0.0575	0.0621	0.0726	0.0870	0.0543	0.0586	0.0685	0.0822
70	0.0552	0.0596	0.0695	0.0831	0.0537	0.0580	0.0676	0.0809	0.0523	0.0564	0.0658	0.0787	0.0510	0.0550	0.0601	0.0768	0.0484	0.0523	0.0610	0.0730
80	0.0492	0.0530	0.0617	0.0737	0.0480	0.0517	0.0602	0.0720	0.0469	0.0505	0.0588	0.0703	0.0458	0.0494	0.0505	0.0687	0.0438	0.0472	0.0550	0.0658
90	0.0444	0.0478	0.0555	0.0663	0.0434	0.0468	0.0543	0.0649	0.0425	0.0458	0.0532	0.0635	0.0417	0.0448	0.0521	0.0623	0.0400	0.0431	0.0501	0.0598
100	0.0405	0.0435	0.0505	0.0602	0.0397	0.0427	0.0496	0.0591	0.0389	0.0419	0.0486	0.0580	0.0382	0.0411	0.0477	0.0569	0.0368	0.0396	0.0460	0.0549

$v = 10$

m	0				10				20				30				50			
n α	0.100	0.050	0.010	0.001	0.100	0.050	0.010	0.001	0.100	0.050	0.010	0.001	0.100	0.050	0.010	0.001	0.100	0.050	0.010	0.001
3	0.5647	0.6025	0.6743	0.7508	0.4410	0.4752	0.5429	0.6199	0.3609	0.3912	0.4524	0.5247	0.3052	0.3321	0.3871	0.4537	0.2330	0.2547	0.2999	0.3560
4	0.4557	0.4884	0.5536	0.6285	0.3738	0.4030	0.4623	0.5328	0.3166	0.3426	0.3961	0.4610	0.2745	0.2978	0.3462	0.4061	0.2167	0.2358	0.2763	0.3271
5	0.3835	0.4118	0.4697	0.5389	0.3252	0.3505	0.4029	0.4667	0.2822	0.3049	0.3524	0.4114	0.2491	0.2697	0.3131	0.3674	0.2018	0.2190	0.2557	0.3025
6	0.3321	0.3568	0.4084	0.4716	0.2883	0.3106	0.3575	0.4159	0.2547	0.2749	0.3177	0.3714	0.2280	0.2465	0.2858	0.3354	0.1886	0.2043	0.2379	0.2810
7	0.2934	0.3154	0.3616	0.4193	0.2593	0.2792	0.3215	0.3749	0.2323	0.2505	0.2893	0.3388	0.2103	0.2271	0.2629	0.3088	0.1768	0.1913	0.2223	0.2624
8	0.2632	0.2829	0.3248	0.3775	0.2358	0.2539	0.2924	0.3413	0.2136	0.2302	0.2657	0.3115	0.1952	0.2105	0.2435	0.2861	0.1664	0.1798	0.2086	0.2461
9	0.2390	0.2568	0.2950	0.3437	0.2165	0.2329	0.2682	0.3135	0.1978	0.2131	0.2458	0.2881	0.1822	0.1963	0.2269	0.2665	0.1572	0.1696	0.1966	0.2317
10	0.2190	0.2353	0.2704	0.3154	0.2002	0.2153	0.2479	0.2900	0.1844	0.1984	0.2288	0.2683	0.1708	0.1840	0.2125	0.2495	0.1489	0.1606	0.1858	0.2190
11	0.2023	0.2173	0.2497	0.2915	0.1863	0.2003	0.2305	0.2698	0.1727	0.1857	0.2141	0.2510	0.1609	0.1731	0.1998	0.2347	0.1415	0.1524	0.1763	0.2076
12	0.1881	0.2020	0.2321	0.2711	0.1743	0.1873	0.2155	0.2523	0.1624	0.1746	0.2012	0.2359	0.1520	0.1635	0.1886	0.2216	0.1348	0.1451	0.1676	0.1972
13	0.1759	0.1888	0.2169	0.2535	0.1639	0.1760	0.2024	0.2369	0.1534	0.1648	0.1898	0.2226	0.1442	0.1550	0.1786	0.2098	0.1287	0.1385	0.1598	0.1881
14	0.1652	0.1773	0.2036	0.2383	0.1546	0.1660	0.1909	0.2236	0.1454	0.1561	0.1797	0.2107	0.1371	0.1473	0.1697	0.1992	0.1232	0.1324	0.1527	0.1797
15	0.1558	0.1671	0.1919	0.2246	0.1465	0.1572	0.1806	0.2116	0.1382	0.1483	0.1706	0.2002	0.1308	0.1404	0.1616	0.1897	0.1181	0.1269	0.1463	0.1719
20	0.1218	0.1305	0.1496	0.1750	0.1162	0.1245	0.1428	0.1672	0.1111	0.1190	0.1366	0.1600	0.1064	0.1140	0.1309	0.1536	0.0981	0.1052	0.1209	0.1418
25	0.1004	0.1074	0.1230	0.1437	0.0967	0.1034	0.1184	0.1385	0.0932	0.0997	0.1142	0.1336	0.0899	0.0962	0.1103	0.1291	0.0841	0.0900	0.1032	0.1208
30	0.0856	0.0915	0.1046	0.1222	0.0829	0.0887	0.1013	0.1185	0.0804	0.0860	0.0983	0.1149	0.0780	0.0834	0.0954	0.1116	0.0737	0.0788	0.0901	0.1055
35	0.0748	0.0799	0.0912	0.1064	0.0728	0.0777	0.0887	0.1036	0.0708	0.0757	0.0864	0.1009	0.0690	0.0737	0.0842	0.0984	0.0656	0.0701	0.0801	0.0936
40	0.0665	0.0709	0.0809	0.0944	0.0649	0.0693	0.0789	0.0922	0.0634	0.0676	0.0771	0.0900	0.0619	0.0661	0.0754	0.0880	0.0592	0.0632	0.0721	0.0842
45	0.0599	0.0639	0.0727	0.0848	0.0586	0.0625	0.0712	0.0830	0.0574	0.0612	0.0697	0.0813	0.0562	0.0600	0.0683	0.0796	0.0540	0.0576	0.0656	0.0765
50	0.0546	0.0581	0.0661	0.0770	0.0535	0.0570	0.0649	0.0756	0.0525	0.0559	0.0636	0.0741	0.0515	0.0549	0.0625	0.0728	0.0497	0.0529	0.0602	0.0702
60	0.0464	0.0494	0.0561	0.0652	0.0456	0.0486	0.0552	0.0641	0.0449	0.0478	0.0543	0.0631	0.0442	0.0471	0.0534	0.0621	0.0429	0.0456	0.0518	0.0603
70	0.0404	0.0430	0.0487	0.0566	0.0399	0.0424	0.0480	0.0558	0.0393	0.0418	0.0474	0.0550	0.0388	0.0412	0.0467	0.0543	0.0378	0.0401	0.0455	0.0529
80	0.0359	0.0381	0.0431	0.0500	0.0354	0.0376	0.0426	0.0494	0.0350	0.0372	0.0421	0.0488	0.0346	0.0367	0.0416	0.0482	0.0338	0.0359	0.0406	0.0471
90	0.0323	0.0342	0.0387	0.0449	0.0319	0.0339	0.0383	0.0444	0.0316	0.0335	0.0379	0.0439	0.0312	0.0331	0.0375	0.0435	0.0306	0.0325	0.0367	0.0425
100	0.0293	0.0311	0.0351	0.0407	0.0290	0.0308	0.0348	0.0403	0.0288	0.0305	0.0345	0.0399	0.0285	0.0302	0.0341	0.0395	0.0279	0.0296	0.0335	0.0388

$v = 15$

n \ m	0				10				20				30				50			
α	0.100	0.050	0.010	0.001	0.100	0.050	0.010	0.001	0.100	0.050	0.010	0.001	0.100	0.050	0.010	0.001	0.100	0.050	0.010	0.001
3	0.5222	0.5536	0.6145	0.6825	0.4376	0.4665	0.5239	0.5902	0.3763	0.4026	0.4557	0.5186	0.3298	0.3539	0.4028	0.4616	0.2644	0.2847	0.3265	0.3779
4	0.4165	0.4430	0.4964	0.5594	0.3623	0.3865	0.4359	0.4954	0.3204	0.3426	0.3883	0.4439	0.2872	0.3076	0.3490	0.4022	0.2377	0.2553	0.2920	0.3378
5	0.3478	0.3703	0.4168	0.4733	0.3099	0.3307	0.3738	0.4267	0.2795	0.2986	0.3386	0.3881	0.2544	0.2722	0.3095	0.3561	0.2157	0.2312	0.2640	0.3052
6	0.2993	0.3189	0.3597	0.4104	0.2714	0.2895	0.3275	0.3749	0.2482	0.2650	0.3005	0.3452	0.2286	0.2443	0.2776	0.3198	0.1974	0.2113	0.2409	0.2784
7	0.2632	0.2804	0.3167	0.3623	0.2417	0.2577	0.2917	0.3347	0.2234	0.2384	0.2703	0.3110	0.2077	0.2218	0.2519	0.2902	0.1820	0.1946	0.2215	0.2561
8	0.2352	0.2506	0.2832	0.3246	0.2181	0.2325	0.2632	0.3024	0.2033	0.2169	0.2458	0.2826	0.1904	0.2032	0.2306	0.2658	0.1689	0.1804	0.2051	0.2371
9	0.2129	0.2267	0.2563	0.2941	0.1989	0.2120	0.2399	0.2758	0.1867	0.1990	0.2255	0.2597	0.1758	0.1875	0.2121	0.2452	0.1575	0.1682	0.1911	0.2207
10	0.1946	0.2072	0.2342	0.2690	0.1830	0.1949	0.2205	0.2537	0.1726	0.1840	0.2084	0.2400	0.1634	0.1742	0.1979	0.2277	0.1477	0.1575	0.1788	0.2065
11	0.1793	0.1908	0.2157	0.2480	0.1695	0.1804	0.2042	0.2349	0.1607	0.1711	0.1938	0.2232	0.1527	0.1627	0.1844	0.2127	0.1390	0.1482	0.1681	0.1941
12	0.1663	0.1770	0.2000	0.2300	0.1579	0.1681	0.1901	0.2187	0.1503	0.1600	0.1811	0.2086	0.1434	0.1527	0.1720	0.1994	0.1313	0.1399	0.1586	0.1831
13	0.1552	0.1651	0.1865	0.2146	0.1479	0.1574	0.1779	0.2048	0.1413	0.1503	0.1701	0.1959	0.1352	0.1439	0.1625	0.1878	0.1245	0.1326	0.1502	0.1733
14	0.1455	0.1547	0.1748	0.2012	0.1391	0.1480	0.1673	0.1925	0.1333	0.1418	0.1603	0.1848	0.1279	0.1361	0.1540	0.1775	0.1184	0.1260	0.1426	0.1645
15	0.1370	0.1457	0.1645	0.1891	0.1314	0.1397	0.1578	0.1816	0.1262	0.1342	0.1517	0.1747	0.1214	0.1291	0.1461	0.1682	0.1128	0.1200	0.1358	0.1566
20	0.1065	0.1130	0.1274	0.1465	0.1031	0.1095	0.1235	0.1421	0.1000	0.1062	0.1197	0.1378	0.0970	0.1030	0.1160	0.1338	0.0916	0.0973	0.1098	0.1264
25	0.0874	0.0927	0.1043	0.1199	0.0852	0.0903	0.1017	0.1168	0.0830	0.0881	0.0992	0.1140	0.0810	0.0860	0.0961	0.1113	0.0773	0.0820	0.0924	0.1063
30	0.0743	0.0787	0.0885	0.1016	0.0727	0.0770	0.0866	0.0994	0.0712	0.0754	0.0848	0.0974	0.0697	0.0739	0.0830	0.0954	0.0670	0.0710	0.0798	0.0917
35	0.0647	0.0685	0.0769	0.0882	0.0635	0.0672	0.0755	0.0866	0.0624	0.0660	0.0741	0.0851	0.0613	0.0649	0.0721	0.0836	0.0592	0.0626	0.0704	0.0807
40	0.0574	0.0607	0.0681	0.0780	0.0565	0.0597	0.0670	0.0768	0.0556	0.0588	0.0659	0.0756	0.0547	0.0579	0.0640	0.0744	0.0530	0.0561	0.0629	0.0721
45	0.0516	0.0546	0.0611	0.0700	0.0509	0.0538	0.0602	0.0690	0.0501	0.0530	0.0594	0.0681	0.0494	0.0523	0.0581	0.0671	0.0481	0.0508	0.0570	0.0653
50	0.0469	0.0496	0.0555	0.0635	0.0463	0.0489	0.0548	0.0627	0.0457	0.0483	0.0541	0.0619	0.0451	0.0477	0.0531	0.0611	0.0440	0.0465	0.0521	0.0596
60	0.0398	0.0420	0.0469	0.0536	0.0393	0.0415	0.0464	0.0530	0.0389	0.0411	0.0459	0.0525	0.0385	0.0406	0.0450	0.0519	0.0377	0.0398	0.0445	0.0508
70	0.0346	0.0365	0.0407	0.0464	0.0342	0.0361	0.0403	0.0460	0.0339	0.0358	0.0399	0.0456	0.0336	0.0355	0.0390	0.0452	0.0330	0.0348	0.0389	0.0444
80	0.0306	0.0323	0.0360	0.0410	0.0304	0.0320	0.0357	0.0407	0.0301	0.0317	0.0354	0.0403	0.0299	0.0315	0.0350	0.0400	0.0294	0.0310	0.0345	0.0394
90	0.0275	0.0290	0.0322	0.0367	0.0273	0.0287	0.0320	0.0365	0.0271	0.0285	0.0318	0.0362	0.0269	0.0283	0.0310	0.0359	0.0265	0.0279	0.0311	0.0355
100	0.0250	0.0263	0.0292	0.0333	0.0248	0.0261	0.0290	0.0331	0.0246	0.0259	0.0289	0.0329	0.0245	0.0258	0.0280	0.0327	0.0242	0.0254	0.0283	0.0322

$v = 20$

m	0				10				20				30				50			
n \ α	0.100	0.050	0.010	0.001	0.100	0.050	0.010	0.001	0.100	0.050	0.010	0.001	0.100	0.050	0.010	0.001	0.100	0.050	0.010	0.001
3	0.4966	0.5239	0.5775	0.6385	0.4327	0.4582	0.5089	0.5682	0.3832	0.4068	0.4543	0.5106	0.3437	0.3656	0.4101	0.4633	0.2848	0.3039	0.3429	0.3905
4	0.3933	0.4159	0.4620	0.5169	0.3530	0.3741	0.4173	0.4694	0.3202	0.3399	0.3804	0.4296	0.2929	0.3113	0.3493	0.3960	0.2502	0.2665	0.3002	0.3422
5	0.3267	0.3459	0.3855	0.4340	0.2990	0.3170	0.3542	0.4001	0.2756	0.2925	0.3275	0.3710	0.2556	0.2714	0.3046	0.3459	0.2231	0.2373	0.2670	0.3044
6	0.2802	0.2967	0.3312	0.3742	0.2599	0.2754	0.3081	0.3489	0.2423	0.2570	0.2879	0.3266	0.2270	0.2409	0.2702	0.3072	0.2014	0.2140	0.2406	0.2744
7	0.2457	0.2601	0.2907	0.3291	0.2301	0.2438	0.2728	0.3095	0.2164	0.2295	0.2570	0.2920	0.2043	0.2167	0.2430	0.2763	0.1837	0.1949	0.2190	0.2497
8	0.2190	0.2319	0.2592	0.2939	0.2067	0.2190	0.2450	0.2783	0.1958	0.2074	0.2323	0.2641	0.1859	0.1970	0.2208	0.2513	0.1688	0.1791	0.2010	0.2292
9	0.1978	0.2094	0.2340	0.2656	0.1878	0.1989	0.2225	0.2528	0.1788	0.1894	0.2120	0.2412	0.1706	0.1808	0.2025	0.2304	0.1563	0.1657	0.1858	0.2119
10	0.1805	0.1910	0.2135	0.2424	0.1722	0.1823	0.2039	0.2317	0.1647	0.1743	0.1951	0.2220	0.1578	0.1671	0.1871	0.2129	0.1456	0.1542	0.1729	0.1970
11	0.1660	0.1756	0.1963	0.2230	0.1591	0.1683	0.1882	0.2140	0.1527	0.1616	0.1808	0.2056	0.1468	0.1554	0.1739	0.1980	0.1362	0.1443	0.1616	0.1841
12	0.1538	0.1627	0.1818	0.2065	0.1479	0.1564	0.1749	0.1988	0.1424	0.1506	0.1685	0.1916	0.1373	0.1452	0.1625	0.1850	0.1281	0.1356	0.1518	0.1728
13	0.1433	0.1516	0.1693	0.1924	0.1382	0.1461	0.1633	0.1858	0.1334	0.1411	0.1577	0.1794	0.1290	0.1364	0.1525	0.1736	0.1209	0.1279	0.1431	0.1630
14	0.1343	0.1419	0.1585	0.1801	0.1298	0.1372	0.1533	0.1743	0.1256	0.1328	0.1484	0.1687	0.1216	0.1286	0.1438	0.1636	0.1145	0.1211	0.1354	0.1542
15	0.1263	0.1335	0.1490	0.1694	0.1223	0.1293	0.1444	0.1642	0.1186	0.1254	0.1401	0.1593	0.1151	0.1217	0.1360	0.1548	0.1087	0.1149	0.1285	0.1462
20	0.0977	0.1032	0.1150	0.1306	0.0954	0.1007	0.1123	0.1276	0.0932	0.0984	0.1097	0.1246	0.0911	0.0961	0.1072	0.1218	0.0871	0.0920	0.1026	0.1166
25	0.0800	0.0843	0.0939	0.1065	0.0784	0.0827	0.0921	0.1045	0.0770	0.0812	0.0904	0.1025	0.0755	0.0797	0.0887	0.1007	0.0728	0.0768	0.0855	0.0972
30	0.0678	0.0715	0.0794	0.0901	0.0668	0.0703	0.0782	0.0887	0.0657	0.0692	0.0769	0.0873	0.0647	0.0681	0.0757	0.0859	0.0627	0.0661	0.0735	0.0834
35	0.0590	0.0621	0.0690	0.0781	0.0582	0.0612	0.0680	0.0771	0.0574	0.0604	0.0671	0.0760	0.0566	0.0596	0.0662	0.0750	0.0551	0.0580	0.0644	0.0730
40	0.0522	0.0550	0.0610	0.0690	0.0516	0.0543	0.0602	0.0682	0.0510	0.0536	0.0595	0.0674	0.0504	0.0530	0.0588	0.0666	0.0492	0.0518	0.0574	0.0651
45	0.0469	0.0493	0.0547	0.0619	0.0464	0.0488	0.0541	0.0612	0.0459	0.0483	0.0535	0.0605	0.0454	0.0478	0.0529	0.0599	0.0445	0.0468	0.0518	0.0587
50	0.0426	0.0448	0.0496	0.0561	0.0422	0.0444	0.0491	0.0555	0.0418	0.0439	0.0486	0.0550	0.0414	0.0435	0.0482	0.0545	0.0406	0.0427	0.0473	0.0535
60	0.0361	0.0379	0.0419	0.0473	0.0358	0.0376	0.0415	0.0469	0.0355	0.0373	0.0412	0.0465	0.0352	0.0369	0.0409	0.0461	0.0346	0.0364	0.0402	0.0454
70	0.0313	0.0328	0.0363	0.0409	0.0311	0.0326	0.0360	0.0406	0.0309	0.0324	0.0358	0.0403	0.0306	0.0322	0.0355	0.0401	0.0302	0.0317	0.0350	0.0395
80	0.0277	0.0290	0.0320	0.0361	0.0275	0.0288	0.0318	0.0359	0.0273	0.0287	0.0316	0.0356	0.0272	0.0285	0.0314	0.0354	0.0268	0.0282	0.0311	0.0350
90	0.0248	0.0260	0.0287	0.0323	0.0247	0.0259	0.0285	0.0321	0.0246	0.0257	0.0284	0.0320	0.0244	0.0256	0.0282	0.0318	0.0242	0.0253	0.0279	0.0314
100	0.0225	0.0236	0.0260	0.0292	0.0224	0.0235	0.0259	0.0291	0.0223	0.0234	0.0257	0.0290	0.0222	0.0233	0.0256	0.0288	0.0220	0.0230	0.0254	0.0285

$v = 30$

n \ m, α	0: 0.100	0: 0.050	0: 0.010	0: 0.001	10: 0.100	10: 0.050	10: 0.010	10: 0.001	20: 0.100	20: 0.050	20: 0.010	20: 0.001	30: 0.100	30: 0.050	30: 0.010	30: 0.001	50: 0.100	50: 0.050	50: 0.010	50: 0.001
3	0.4662	0.4885	0.5327	0.5840	0.4236	0.4447	0.4869	0.5364	0.3880	0.4080	0.4482	0.4959	0.3579	0.3768	0.4151	0.4608	0.3097	0.3267	0.3614	0.4032
4	0.3659	0.3842	0.4213	0.4658	0.3397	0.3570	0.3923	0.4351	0.3169	0.3334	0.3671	0.4081	0.2970	0.3126	0.3448	0.3842	0.2638	0.2780	0.3075	0.3437
5	0.3022	0.3174	0.3489	0.3876	0.2844	0.2989	0.3290	0.3661	0.2685	0.2824	0.3112	0.3470	0.2543	0.2676	0.2953	0.3296	0.2300	0.2422	0.2678	0.2997
6	0.2579	0.2709	0.2982	0.3322	0.2450	0.2575	0.2836	0.3164	0.2333	0.2453	0.2705	0.3020	0.2227	0.2342	0.2584	0.2888	0.2041	0.2148	0.2373	0.2657
7	0.2253	0.2367	0.2606	0.2907	0.2155	0.2265	0.2495	0.2785	0.2065	0.2171	0.2393	0.2675	0.1983	0.2084	0.2300	0.2572	0.1836	0.1931	0.2132	0.2388
8	0.2003	0.2103	0.2316	0.2588	0.1926	0.2023	0.2229	0.2490	0.1854	0.1948	0.2148	0.2402	0.1788	0.1879	0.2073	0.2319	0.1669	0.1755	0.1937	0.2169
9	0.1804	0.1894	0.2086	0.2331	0.1742	0.1829	0.2015	0.2253	0.1684	0.1769	0.1949	0.2180	0.1630	0.1712	0.1887	0.2112	0.1531	0.1609	0.1775	0.1987
10	0.1642	0.1724	0.1898	0.2122	0.1591	0.1670	0.1840	0.2058	0.1543	0.1620	0.1785	0.1997	0.1498	0.1573	0.1733	0.1940	0.1414	0.1486	0.1638	0.1835
11	0.1508	0.1583	0.1742	0.1948	0.1465	0.1539	0.1693	0.1894	0.1424	0.1495	0.1647	0.1843	0.1386	0.1455	0.1603	0.1794	0.1315	0.1381	0.1522	0.1704
12	0.1395	0.1463	0.1611	0.1801	0.1358	0.1425	0.1569	0.1755	0.1323	0.1389	0.1529	0.1711	0.1290	0.1354	0.1491	0.1669	0.1229	0.1290	0.1421	0.1592
13	0.1298	0.1361	0.1498	0.1675	0.1266	0.1328	0.1462	0.1636	0.1236	0.1297	0.1427	0.1597	0.1208	0.1267	0.1395	0.1560	0.1154	0.1211	0.1333	0.1493
14	0.1214	0.1273	0.1400	0.1566	0.1186	0.1244	0.1369	0.1531	0.1160	0.1217	0.1339	0.1497	0.1135	0.1190	0.1310	0.1466	0.1088	0.1141	0.1256	0.1406
15	0.1141	0.1196	0.1315	0.1470	0.1116	0.1170	0.1287	0.1439	0.1093	0.1146	0.1261	0.1410	0.1071	0.1123	0.1235	0.1382	0.1029	0.1079	0.1187	0.1329
20	0.0878	0.0920	0.1010	0.1129	0.0864	0.0905	0.0994	0.1111	0.0850	0.0891	0.0978	0.1094	0.0837	0.0877	0.0963	0.1077	0.0812	0.0850	0.0934	0.1045
25	0.0716	0.0749	0.0822	0.0918	0.0707	0.0740	0.0811	0.0906	0.0698	0.0730	0.0801	0.0894	0.0689	0.0721	0.0791	0.0883	0.0672	0.0703	0.0772	0.0862
30	0.0606	0.0633	0.0694	0.0774	0.0599	0.0627	0.0686	0.0766	0.0593	0.0620	0.0679	0.0757	0.0587	0.0613	0.0672	0.0749	0.0574	0.0601	0.0658	0.0734
35	0.0526	0.0549	0.0601	0.0670	0.0521	0.0544	0.0596	0.0664	0.0516	0.0539	0.0590	0.0658	0.0511	0.0534	0.0585	0.0652	0.0502	0.0525	0.0574	0.0640
40	0.0465	0.0485	0.0531	0.0591	0.0461	0.0481	0.0526	0.0586	0.0457	0.0477	0.0522	0.0582	0.0453	0.0474	0.0518	0.0577	0.0446	0.0466	0.0510	0.0568
45	0.0417	0.0435	0.0475	0.0529	0.0414	0.0432	0.0472	0.0525	0.0411	0.0429	0.0468	0.0521	0.0408	0.0426	0.0465	0.0518	0.0402	0.0420	0.0459	0.0510
50	0.0378	0.0394	0.0431	0.0479	0.0375	0.0392	0.0428	0.0476	0.0373	0.0389	0.0425	0.0473	0.0371	0.0387	0.0422	0.0470	0.0366	0.0382	0.0417	0.0464
60	0.0319	0.0333	0.0363	0.0403	0.0317	0.0331	0.0361	0.0401	0.0316	0.0329	0.0359	0.0399	0.0314	0.0327	0.0357	0.0397	0.0310	0.0324	0.0353	0.0392
70	0.0276	0.0288	0.0314	0.0349	0.0275	0.0287	0.0312	0.0347	0.0274	0.0285	0.0311	0.0345	0.0273	0.0284	0.0309	0.0344	0.0270	0.0281	0.0307	0.0340
80	0.0244	0.0254	0.0277	0.0307	0.0243	0.0253	0.0276	0.0306	0.0242	0.0252	0.0274	0.0305	0.0241	0.0251	0.0273	0.0303	0.0239	0.0249	0.0271	0.0301
90	0.0219	0.0228	0.0248	0.0274	0.0218	0.0227	0.0247	0.0273	0.0217	0.0226	0.0246	0.0273	0.0216	0.0225	0.0245	0.0271	0.0215	0.0223	0.0243	0.0270
100	0.0198	0.0206	0.0224	0.0248	0.0197	0.0205	0.0223	0.0247	0.0197	0.0205	0.0223	0.0247	0.0196	0.0204	0.0222	0.0246	0.0195	0.0203	0.0220	0.0244

$v = 50$

m	0				10				20				30				50			
n \ α	0.100	0.050	0.010	0.001	0.100	0.050	0.010	0.001	0.100	0.050	0.010	0.001	0.100	0.050	0.010	0.001	0.100	0.050	0.010	0.001
3	0.4358	0.4530	0.4872	0.5277	0.4105	0.4270	0.4601	0.4993	0.3879	0.4038	0.4359	0.4740	0.3676	0.3830	0.4140	0.4510	0.3328	0.3471	0.3761	0.4111
4	0.3389	0.3527	0.3808	0.4150	0.3236	0.3370	0.3642	0.3974	0.3097	0.3226	0.3490	0.3811	0.2969	0.3093	0.3349	0.3661	0.2742	0.2859	0.3100	0.3395
5	0.2780	0.2895	0.3131	0.3422	0.2678	0.2789	0.3018	0.3301	0.2583	0.2691	0.2914	0.3189	0.2495	0.2599	0.2816	0.3085	0.2335	0.2434	0.2639	0.2894
6	0.2362	0.2459	0.2661	0.2913	0.2288	0.2383	0.2580	0.2826	0.2219	0.2311	0.2504	0.2744	0.2155	0.2244	0.2432	0.2666	0.2036	0.2121	0.2300	0.2524
7	0.2055	0.2139	0.2316	0.2538	0.2000	0.2082	0.2255	0.2472	0.1948	0.2028	0.2197	0.2409	0.1898	0.1977	0.2142	0.2349	0.1806	0.1881	0.2039	0.2239
8	0.1821	0.1895	0.2052	0.2249	0.1778	0.1850	0.2004	0.2198	0.1737	0.1808	0.1958	0.2148	0.1698	0.1767	0.1914	0.2101	0.1624	0.1691	0.1833	0.2011
9	0.1636	0.1702	0.1842	0.2021	0.1601	0.1666	0.1804	0.1980	0.1568	0.1632	0.1767	0.1939	0.1536	0.1599	0.1732	0.1901	0.1476	0.1537	0.1665	0.1828
10	0.1486	0.1546	0.1673	0.1835	0.1457	0.1516	0.1641	0.1801	0.1430	0.1488	0.1611	0.1767	0.1404	0.1460	0.1581	0.1736	0.1354	0.1409	0.1526	0.1675
11	0.1362	0.1416	0.1532	0.1681	0.1338	0.1391	0.1506	0.1652	0.1315	0.1368	0.1480	0.1624	0.1293	0.1345	0.1456	0.1597	0.1251	0.1301	0.1408	0.1546
12	0.1257	0.1307	0.1414	0.1551	0.1237	0.1286	0.1392	0.1527	0.1217	0.1266	0.1370	0.1503	0.1198	0.1246	0.1349	0.1481	0.1162	0.1209	0.1308	0.1436
13	0.1168	0.1214	0.1313	0.1440	0.1150	0.1196	0.1294	0.1419	0.1134	0.1179	0.1275	0.1399	0.1117	0.1162	0.1257	0.1379	0.1086	0.1129	0.1222	0.1341
14	0.1091	0.1134	0.1226	0.1344	0.1076	0.1118	0.1209	0.1326	0.1061	0.1103	0.1193	0.1309	0.1047	0.1088	0.1177	0.1291	0.1019	0.1060	0.1146	0.1258
15	0.1023	0.1063	0.1150	0.1261	0.1010	0.1050	0.1135	0.1245	0.0997	0.1036	0.1121	0.1229	0.0985	0.1023	0.1106	0.1214	0.0961	0.0998	0.1079	0.1185
20	0.0784	0.0814	0.0879	0.0963	0.0776	0.0806	0.0870	0.0954	0.0769	0.0798	0.0862	0.0945	0.0761	0.0791	0.0854	0.0936	0.0747	0.0776	0.0838	0.0918
25	0.0637	0.0661	0.0713	0.0781	0.0632	0.0656	0.0707	0.0775	0.0627	0.0650	0.0702	0.0768	0.0622	0.0645	0.0696	0.0763	0.0613	0.0636	0.0686	0.0751
30	0.0537	0.0557	0.0600	0.0657	0.0534	0.0553	0.0596	0.0653	0.0530	0.0550	0.0592	0.0648	0.0527	0.0546	0.0589	0.0644	0.0520	0.0539	0.0581	0.0636
35	0.0465	0.0482	0.0519	0.0568	0.0462	0.0479	0.0516	0.0564	0.0460	0.0476	0.0513	0.0561	0.0457	0.0474	0.0510	0.0558	0.0452	0.0469	0.0505	0.0552
40	0.0410	0.0425	0.0457	0.0500	0.0408	0.0423	0.0455	0.0497	0.0406	0.0421	0.0453	0.0495	0.0404	0.0419	0.0451	0.0493	0.0400	0.0415	0.0446	0.0488
45	0.0367	0.0380	0.0409	0.0447	0.0366	0.0379	0.0407	0.0445	0.0364	0.0377	0.0405	0.0443	0.0362	0.0375	0.0404	0.0441	0.0359	0.0372	0.0400	0.0437
50	0.0333	0.0344	0.0370	0.0404	0.0331	0.0343	0.0369	0.0402	0.0330	0.0342	0.0367	0.0401	0.0329	0.0340	0.0366	0.0399	0.0326	0.0338	0.0363	0.0396
60	0.0280	0.0290	0.0311	0.0339	0.0279	0.0289	0.0310	0.0338	0.0278	0.0288	0.0309	0.0337	0.0277	0.0287	0.0308	0.0336	0.0275	0.0285	0.0306	0.0334
70	0.0242	0.0250	0.0269	0.0293	0.0241	0.0250	0.0268	0.0292	0.0241	0.0249	0.0267	0.0291	0.0240	0.0248	0.0266	0.0290	0.0239	0.0247	0.0265	0.0289
80	0.0213	0.0221	0.0236	0.0258	0.0213	0.0220	0.0236	0.0257	0.0212	0.0219	0.0235	0.0256	0.0212	0.0219	0.0235	0.0256	0.0211	0.0218	0.0234	0.0255
90	0.0191	0.0197	0.0211	0.0230	0.0190	0.0197	0.0211	0.0230	0.0190	0.0196	0.0210	0.0229	0.0190	0.0196	0.0210	0.0229	0.0189	0.0195	0.0209	0.0228
100	0.0173	0.0178	0.0191	0.0208	0.0172	0.0178	0.0191	0.0208	0.0172	0.0178	0.0190	0.0207	0.0172	0.0177	0.0190	0.0207	0.0171	0.0177	0.0189	0.0206

APPENDIX 10

A single outlier in a two-way factorial experiment

Given X_{ij}, $j = 1$ to C, $i = 1$ to R, a set of data from an $R \times C$ non-replicated factorial design, let e_{ij} be the residual corresponding to X_{ij}. The test satistic is $\max|e_{ij}|/(\Sigma_i \Sigma_j e_{ij}^2)^{1/2}$, and its fractiles may be bracketed by successive partial sums in the Boole expansion.

The table given was computed using the first two terms of the Boole expansion. In most of the cases, either the two approximands bracketing the true fractiles agreed to 3 decimals, or one of them was known to be exact. In the remaining cases, only the conservative fractile given by the first term is listed, and is marked with an asterisk.

When relating these fractiles to the discussion of Chapter 7, one should be aware that the maximum normed residual differs from the Ellenberg statistic of Equation (7.3) in that it has not been divided by $\sqrt{a_{ii}}$ which for this design is $\{(R-1)(C-1)/RC\}^{1/2}$.

There are some minor differences among those values common to both this table and those of Stefansky (1972).

Outside the range of the table, the Bonferroni bound provides an excellent approximation, which is poor only when simultaneously R is large and C very small.

Fractiles of maximum normed residual. $\alpha = 0.10$

	R							
C	3	4	5	6	7	8	9	10
3	0.637	0.625	0.599	0.573	0.548	0.529*	0.511*	0.494*
4		0.594	0.562	0.533	0.508	0.486	0.468*	0.451*
5			0.527	0.498	0.473	0.452	0.433	0.417*
6				0.469	0.444	0.424	0.406	0.390
7					0.421	0.401	0.383	0.369*
8						0.381	0.365	0.350
9							0.349*	0.335*
10								0.321

Fractiles of maximum normed residual. $\alpha = 0.05$

	R							
C	3	4	5	6	7	8	9	10
3	0.648	0.645	0.624	0.600	0.577	0.555	0.537*	0.520*
4		0.621	0.590	0.561	0.535	0.513	0.493	0.476*
5			0.555	0.525	0.499	0.477	0.457	0.440
6				0.495	0.469	0.447	0.428	0.412
7					0.444	0.423	0.405	0.389
8						0.402	0.385	0.369
9							0.368	0.353
10								0.338

Fractiles of maximum normed residual. $\alpha = 0.01$

	R							
C	3	4	5	6	7	8	9	10
3	0.660	0.675	0.664	0.646	0.626	0.606	0.587	0.570
4		0.665	0.640	0.613	0.588	0.565	0.544	0.525
5			0.608	0.578	0.551	0.527	0.506	0.488
6				0.546	0.519	0.495	0.475	0.457
7					0.492	0.469	0.449	0.431
8						0.446	0.426	0.409
9							0.407	0.391
10								0.375

APPENDIX 11

Fractiles of $X_{(n)}$ for the Poisson, binomial and negative binomial distributions

Given $X_1, \ldots, X_n$ drawn from a discrete generalized power series distribution (GPSD), the statistic of interest for testing for slippage to the right is $X_{(n)}$, whose distribution must be conditioned on $T = \Sigma_1^n X_i$ to eliminate the unknown nuisance parameter. The general distribution is given by

$$f_n(x, t) = \Pr[X_{(n)} = x \mid T = t]$$
$$= \sum_{j=1}^{[t/x]} \binom{n}{j} \left\{ \prod_{i=0}^{j-1} h_{n-i}(x, t - ix) \right\} (-1)^j F_{n-j}(x, t - jx)$$

where $F_n(x, t) = \sum_{y=0}^{x} f_n(x, t)$

and $h_n(x, t)$ is:

$$\frac{t!(n-1)^{t-x}}{n^t x!(t-x)!} \qquad \text{if } X_i \sim \text{Poisson } (\lambda);$$

$$\frac{\binom{r}{x}\binom{nr-r}{t-x}}{\binom{nr}{t}} \qquad \text{if } X_i \text{ is binomial } (r, \theta);$$

$$\frac{\binom{-r}{x}\binom{-nr+r}{t-x}}{\binom{-nr}{t}} \qquad \text{if } X_i \text{ is negative binomial } (r, \theta).$$

The fractiles quoted are the smallest x such that $F_n(x, t) \geq 1 - \alpha$.

As r increases, both the binomial and the negative binomial distributions tend towards the Poisson; as the tables suggest, the Poisson theory provides an excellent approximation even for moderately small r. This fact largely eliminates the need for very comprehensive tables for the binomial or negative binomial distribution.

Fractiles for Negative binomial (5, θ)

		T													
n	α	4	6	8	10	12	14	16	18	20	25	30	35	40	50
4	0.10	3	4	5	6	7	8	9	10	11	13	15	18	20	25
	0.05	3	4	6	7	8	9	10	11	12	14	17	19	22	27
	0.01	4	5	6	8	9	10	11	12	14	16	19	22	25	30
6	0.10	3	3	4	5	6	6	7	8	9	10	12	14	15	19
	0.05	3	4	5	6	6	7	8	9	9	11	13	15	17	20
	0.01	4	5	6	7	7	8	9	10	11	13	15	17	19	24
8	0.10	2	3	4	4	5	6	6	7	7	9	10	11	13	16
	0.05	3	3	4	5	6	6	7	7	8	10	11	13	14	17
	0.01	3	4	5	6	7	7	8	9	10	11	13	15	16	20
10	0.10	2	3	3	4	5	5	6	6	7	8	9	10	11	13
	0.05	3	3	4	4	5	6	6	7	7	8	10	11	12	15
	0.01	3	4	5	5	6	7	7	8	9	10	11	13	14	17
12	0.10	2	3	3	4	4	5	5	6	6	7	8	9	10	12
	0.05	2	3	4	4	5	5	6	6	7	8	9	10	11	13
	0.01	3	4	4	5	6	6	7	7	8	9	10	12	13	15
14	0.10	2	3	3	4	4	4	5	5	6	6	7	8	9	11
	0.05	2	3	3	4	4	5	5	6	6	7	8	9	10	12
	0.01	3	4	4	5	5	6	6	7	7	8	10	11	12	14
16	0.10	2	3	3	3	4	4	4	5	5	6	7	8	8	10
	0.05	2	3	3	4	4	5	5	5	6	7	8	8	9	11
	0.01	3	3	4	4	5	5	6	6	7	8	9	10	11	13
18	0.10	2	2	3	3	4	4	4	5	5	6	6	7	8	9
	0.05	2	3	3	4	4	4	5	5	5	6	7	8	9	10
	0.01	3	3	4	4	5	5	6	6	6	7	8	9	10	12
20	0.10	2	2	3	3	3	4	4	4	5	5	6	7	7	9
	0.05	2	3	3	3	4	4	5	5	5	6	7	7	8	10
	0.01	3	3	4	4	5	5	5	6	6	7	8	9	10	11
25	0.10	2	2	3	3	3	3	4	4	4	5	5	6	7	8
	0.05	2	2	3	3	4	4	4	4	5	5	6	7	7	8
	0.01	3	3	3	4	4	5	5	5	6	6	7	8	9	10
30	0.10	2	2	2	3	3	3	3	4	4	5	5	6	6	7
	0.05	2	2	3	3	3	4	4	4	4	5	6	6	7	8
	0.01	2	3	3	4	4	4	5	5	5	6	7	7	8	9
35	0.10	2	2	2	3	3	3	3	4	4	4	5	5	6	7
	0.05	2	2	3	3	3	3	4	4	4	5	5	6	6	7
	0.01	2	3	3	3	4	4	4	5	5	6	6	7	7	8
40	0.10	2	2	2	2	3	3	3	3	4	4	4	5	5	6
	0.05	2	2	2	3	3	3	3	4	4	4	5	5	6	7
	0.01	2	3	3	3	4	4	4	4	5	5	6	6	7	8
50	0.10	2	2	2	2	3	3	3	3	3	4	4	4	5	5
	0.05	2	2	2	3	3	3	3	3	4	4	4	5	5	6
	0.01	2	3	3	3	3	4	4	4	4	5	5	6	6	7

Fractiles for Negative binomial (10, θ)

		T													
n	α	4	6	8	10	12	14	16	18	20	25	30	35	40	50
4	0.10	3	4	5	6	7	7	8	9	10	12	14	16	18	22
	0.05	3	4	5	6	7	8	9	10	11	13	15	17	19	24
	0.01	4	5	6	7	8	9	10	11	12	15	17	20	22	27
6	0.10	3	3	4	5	5	6	7	7	8	9	11	12	14	17
	0.05	3	4	4	5	6	7	7	8	9	10	12	13	15	18
	0.01	3	4	5	6	7	8	9	9	10	12	14	15	17	21
8	0.10	2	3	4	4	5	5	6	6	7	8	9	10	12	14
	0.05	3	3	4	5	5	6	6	7	7	9	10	11	12	15
	0.01	3	4	5	5	6	7	7	8	9	10	12	13	14	17
10	0.10	2	3	3	4	4	5	5	6	6	7	8	9	10	12
	0.05	2	3	4	4	5	5	6	6	7	8	9	10	11	13
	0.01	3	4	4	5	6	6	7	7	8	9	10	11	13	15
12	0.10	2	3	3	4	4	4	5	5	6	6	7	8	9	11
	0.05	2	3	3	4	4	5	5	6	6	7	8	9	10	12
	0.01	3	4	4	5	5	6	6	7	7	8	9	10	11	13
14	0.10	2	3	3	3	4	4	4	5	5	6	7	8	8	10
	0.05	2	3	3	4	4	5	5	5	6	7	7	8	9	11
	0.01	3	3	4	4	5	5	6	6	7	8	9	10	11	12
16	0.10	2	2	3	3	4	4	4	5	5	6	6	7	8	9
	0.05	2	3	3	4	4	4	5	5	5	6	7	8	8	10
	0.01	3	3	4	4	5	5	6	6	6	7	8	9	10	11
18	0.10	2	2	3	3	3	4	4	4	5	5	6	7	7	9
	0.05	2	3	3	3	4	4	4	5	5	6	7	7	8	9
	0.01	3	3	4	4	5	5	5	6	6	7	8	8	9	11
20	0.10	2	2	3	3	3	4	4	4	4	5	6	6	7	8
	0.05	2	3	3	3	4	4	4	5	5	6	6	7	7	9
	0.01	3	3	4	4	4	5	5	5	6	7	7	8	9	10
25	0.10	2	2	2	3	3	3	4	4	4	5	5	6	6	7
	0.05	2	2	3	3	3	4	4	4	4	5	6	6	7	8
	0.01	2	3	3	4	4	4	5	5	5	6	7	7	8	9
30	0.01	2	2	2	3	3	3	3	4	4	4	5	5	6	6
	0.05	2	2	3	3	3	3	4	4	4	5	5	6	6	7
	0.01	2	3	3	4	4	4	4	5	5	6	6	7	7	8
35	0.10	2	2	2	3	3	3	3	3	4	4	4	5	5	6
	0.05	2	2	3	3	3	3	3	4	4	4	5	5	6	7
	0.01	2	3	3	3	4	4	4	4	5	5	6	6	7	8
40	0.10	2	2	2	2	3	3	3	3	3	4	4	5	5	6
	0.05	2	2	2	3	3	3	3	4	4	4	5	5	5	6
	0.01	2	3	3	3	4	4	4	4	4	5	5	6	6	7
50	0.10	2	2	2	2	2	3	3	3	3	4	4	4	5	5
	0.05	2	2	2	3	3	3	3	3	3	4	4	5	5	6
	0.01	2	2	3	3	3	4	4	4	4	5	5	5	6	7

Fractiles for Binomial (5, θ)

		T													
n	α	4	6	8	10	12	14	16	18	20	25	30	35	40	50
4	0.10	3	3	4	4	5	—	—	—	—	—	—	—	—	
	0.05	3	4	4	5	—	—	—	—	—	—	—	—	—	—
	0.01	3	4	5	5	—	—	—	—	—	—	—	—	—	—
6	0.10	2	3	3	4	4	4	5	5	—	—	—	—	—	—
	0.05	3	3	4	4	4	5	5	5	—	—	—	—	—	—
	0.01	3	4	4	5	5	5	5	—	—	—	—	—	—	—
8	0.10	2	3	3	3	4	4	4	5	5	5	—	—	—	—
	0.05	2	3	3	4	4	4	5	5	5	5	—	—	—	—
	0.05	2	3	3	4	4	4	5	5	5	5	—	—	—	—
	0.01	3	4	4	4	4	5	5	5	5	5	—	—	—	—
10	0.10	2	2	3	3	3	4	4	4	4	5	5	5	—	—
	0.05	2	3	3	3	4	4	4	4	5	5	5	—	—	—
	0.01	3	3	4	4	4	4	5	5	5	5	5	—	—	—
12	0.10	2	2	3	3	3	4	4	4	4	5	5	5	5	—
	0.05	2	3	3	3	4	4	4	4	4	5	5	5	5	—
	0.01	3	3	3	4	4	4	4	5	5	5	5	5	—	—
14	0.10	2	2	3	3	3	3	4	4	4	4	5	5	5	5
	0.05	2	2	3	3	3	4	4	4	4	5	5	5	5	—
	0.01	3	3	3	4	4	4	4	4	5	5	5	5	5	—
16	0.10	2	2	2	3	3	3	3	4	4	4	4	5	5	5
	0.05	2	2	3	3	3	3	4	4	4	4	5	5	5	5
	0.01	2	3	3	3	4	4	4	4	5	5	5	5	5	5
18	0.10	2	2	2	3	3	3	3	3	4	4	4	5	5	5
	0.05	2	2	3	3	3	3	4	4	4	4	5	5	5	5
	0.01	2	3	3	3	4	4	4	4	4	5	5	5	5	5
20	0.10	2	2	2	3	3	3	3	3	4	4	4	4	5	5
	0.05	2	2	3	3	3	3	3	4	4	4	4	5	5	5
	0.01	2	3	3	3	4	4	4	4	4	5	5	5	5	5
25	0.10	2	2	2	2	3	3	3	3	3	4	4	4	4	5
	0.05	2	2	2	3	3	3	3	3	4	4	4	4	5	5
	0.01	2	3	3	3	3	4	4	4	4	4	5	5	5	5
30	0.10	2	2	2	2	3	3	3	3	3	3	4	4	4	5
	0.05	2	2	2	3	3	3	3	3	3	4	4	4	4	5
	0.01	2	3	3	3	3	3	4	4	4	4	4	5	5	5
35	0.10	2	2	2	2	2	3	3	3	3	3	4	4	4	4
	0.05	2	2	2	2	3	3	3	3	3	4	4	4	4	5
	0.01	2	2	3	3	3	3	3	4	4	4	4	4	5	5
40	0.10	2	2	2	2	2	2	3	3	3	3	3	4	4	4
	0.05	2	2	2	2	3	3	3	3	3	3	4	4	4	4
	0.01	2	2	3	3	3	3	3	3	4	4	4	4	5	5
50	0.10	1	2	2	2	2	2	2	3	3	3	3	3	4	4
	0.05	2	2	2	2	2	3	3	3	3	3	3	4	4	4
	0.01	2	2	3	3	3	3	3	3	3	4	4	4	4	5

Fractiles for Binomial (10, θ)

		T													
n	α	4	6	8	10	12	14	16	18	20	25	30	35	40	50
4	0.10	3	4	4	5	6	6	7	7	8	—	—	—	—	—
	0.05	3	4	5	5	6	6	7	8	8	—	—	—	—	—
	0.01	3	4	5	6	7	7	8	8	9	—	—	—	—	—
6	0.10	2	3	4	4	5	5	5	6	6	7	8	9	—	—
	0.05	3	3	4	4	5	5	6	6	7	8	8	9	—	—
	0.01	3	4	5	5	6	6	7	7	7	8	9	—	—	—
8	0.10	2	3	3	4	4	4	5	5	6	6	7	8	8	10
	0.05	2	3	4	4	4	5	5	6	6	7	7	8	9	—
	0.01	3	4	4	5	5	6	6	6	7	7	8	9	9	—
10	0.10	2	3	3	3	4	4	4	5	5	6	6	7	7	8
	0.05	2	3	3	4	4	4	5	5	5	6	7	7	8	9
	0.01	3	3	4	4	5	5	5	6	6	7	7	8	8	9
12	0.10	2	2	3	3	4	4	4	4	5	5	6	6	7	8
	0.05	2	3	3	3	4	4	4	5	5	6	6	7	7	8
	0.01	3	3	4	4	4	5	5	5	6	6	7	7	8	9
14	0.10	2	2	3	3	3	4	4	4	4	5	5	6	6	7
	0.05	2	3	3	3	4	4	4	4	5	5	6	6	7	8
	0.01	3	3	4	4	4	5	5	5	5	6	7	7	7	8
16	0.10	2	2	3	3	3	3	4	4	4	5	5	6	6	7
	0.05	2	3	3	3	3	4	4	4	4	5	6	6	6	7
	0.01	3	3	3	4	4	4	5	5	5	6	6	7	7	8
18	0.10	2	2	3	3	3	3	4	4	4	4	5	5	6	6
	0.05	2	2	3	3	3	4	4	4	4	5	5	6	6	7
	0.01	2	3	3	4	4	4	5	5	5	6	6	6	7	8
20	0.10	2	2	2	3	3	3	3	4	4	4	5	5	5	6
	0.05	2	2	3	3	3	4	4	4	4	5	5	5	6	7
	0.01	2	3	3	4	4	4	4	5	5	5	6	6	7	7
25	0.10	2	2	2	3	3	3	3	3	4	4	4	5	5	6
	0.05	2	2	3	3	3	3	3	4	4	4	5	5	5	6
	0.01	2	3	3	3	4	4	4	4	4	5	5	6	6	7
30	0.01	2	2	2	2	3	3	3	3	3	4	4	4	5	5
	0.15	2	2	2	3	3	3	3	3	4	4	4	5	5	6
	0.01	2	3	3	3	3	4	4	4	4	5	5	5	6	6
35	0.10	2	2	2	2	3	3	3	3	3	4	4	4	4	5
	0.05	2	2	2	3	3	3	3	3	3	4	4	4	5	5
	0.01	2	3	3	3	3	4	4	4	4	4	5	5	5	6
40	0.10	2	2	2	2	2	3	3	3	3	3	4	4	4	5
	0.15	2	2	2	2	3	3	3	3	3	4	4	4	5	5
	0.01	2	2	3	3	3	3	4	4	4	4	5	5	5	6
50	0.10	2	2	2	2	2	2	3	3	3	3	3	4	4	4
	0.05	2	2	2	2	3	3	3	3	3	3	4	4	4	5
	0.01	2	2	3	3	3	3	3	4	4	4	4	5	5	5

Fractiles for Poisson (θ)

		T													
n	α	4	6	8	10	12	14	16	18	20	25	30	35	40	50
4	0.10	3	4	5	5	6	7	8	8	9	11	12	14	16	19
	0.05	3	4	5	6	7	7	8	9	10	11	13	15	16	20
	0.01	4	5	6	7	8	8	9	10	11	13	15	16	18	22
6	0.10	2	3	4	4	5	6	6	7	7	8	10	11	12	14
	0.05	3	4	4	5	5	6	7	7	8	9	10	12	13	15
	0.01	3	4	5	6	6	7	8	8	9	10	12	13	14	17
8	0.10	2	3	3	4	4	5	5	6	6	7	8	9	10	12
	0.05	3	3	4	4	5	5	6	6	7	8	9	10	11	13
	0.01	3	4	4	5	6	6	7	7	8	9	10	11	12	14
10	0.10	2	3	3	4	4	4	5	5	6	6	7	8	9	10
	0.05	2	3	4	4	4	5	5	6	6	7	8	9	10	11
	0.01	3	4	4	5	5	6	6	7	7	8	9	10	11	13
12	0.10	2	3	3	3	4	4	4	5	5	6	7	7	8	9
	0.05	2	3	3	4	4	5	5	5	6	6	7	8	9	10
	0.01	3	3	4	4	5	5	6	6	6	7	8	9	10	11
14	0.10	2	2	3	3	4	4	4	5	5	5	6	7	7	9
	0.05	2	3	3	4	4	4	5	5	5	6	7	7	8	9
	0.01	3	3	4	4	5	5	5	6	6	7	8	8	9	10
16	0.10	2	2	3	3	3	4	4	4	5	5	6	6	7	8
	0.05	2	3	3	3	4	4	4	5	5	6	6	7	7	9
	0.01	3	3	4	4	4	5	5	5	6	7	7	8	9	10
18	0.10	2	2	3	3	3	4	4	4	4	5	5	6	7	8
	0.05	2	3	3	3	4	4	4	4	5	5	6	6	7	8
	0.01	3	3	4	4	4	5	5	5	6	6	7	7	8	9
20	0.10	2	2	3	3	3	3	4	4	4	5	5	6	6	7
	0.05	2	2	3	3	3	4	4	4	5	5	6	6	7	8
	0.01	2	3	3	4	4	4	5	5	5	6	7	7	8	9
25	0.10	2	2	2	3	3	3	3	4	4	4	5	5	6	6
	0.05	2	2	3	3	3	3	4	4	4	5	5	6	6	7
	0.01	2	3	3	4	4	4	4	5	5	5	6	7	7	8
30	0.10	2	2	2	3	3	3	3	3	4	4	4	5	5	6
	0.05	2	2	3	3	3	3	3	4	4	4	5	5	6	6
	0.01	2	3	3	3	4	4	4	4	5	5	6	6	6	7
35	0.10	2	2	2	2	3	3	3	3	3	4	4	5	5	6
	0.05	2	2	2	3	3	3	3	4	4	4	4	5	5	6
	0.01	2	3	3	3	3	4	4	4	4	5	5	6	6	7
40	0.10	2	2	2	2	3	3	3	3	3	4	4	4	5	5
	0.05	2	2	2	3	3	3	3	3	4	4	4	5	5	6
	0.01	2	3	3	3	3	4	4	4	4	5	5	5	6	6
50	0.10	2	2	2	2	2	3	3	3	3	3	4	4	4	5
	0.05	2	2	2	2	3	3	3	3	3	4	4	4	5	5
	0.01	2	2	3	3	3	3	4	4	4	4	5	5	5	6

Index

see also the Table of Contents

9789401539968

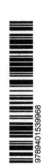